ELECTRICAL POWER SIMPLIFIED

DR. PRASHOBH KARUNAKARAN

AuthorHouse™
1663 Liberty Drive
Bloomington, IN 47403
www.authorhouse.com
Phone: 1 (800) 839-8640

Published by AuthorHouse: 12/31/2015

ISBN: 978-1-5049-6542-2 (sc)
ISBN: 978-1-5049-6543-9 (e)

Library of Congress Control Number: 2015920714

Print information available on the last page.

Any people depicted in stock imagery provided by Thinkstock are models, and such images are being used for illustrative purposes only. Certain stock imagery © Thinkstock.

This book is printed on acid-free paper.

authorHOUSE®

ACKNOWLEDGEMENT

Special thanks must be given to Edi Jungan, K., Harry Ante, Mid Dewang, Phang Su Ling and Ganesa Ramamoorthy for their help in getting some of the materials for this book. Great thanks must also to out to my wife, Sreeja, and children, Prashanth, Shanthi and Arjun for supporting in the writing of this book.

PREFACE

This book simplifies electrical power engineering and provides a working knowledge of the field. Equations are avoided as far as possible. I did a Bachelors in Electrical Engineering and then switched to a Masters in Economics; mainly because the Engineering Department required me to pay fees. The Economics Department offered me full scholarship since I became the top out of 80 students in two economic courses I took in the summer break after securing my engineering degree. I continued acing the Masters subjects and did better than Rich who shared an office room with me at South Dakota State University. But I always knew Rich was much better than me at Economics. I aced economics because of my engineering knowledge of mathematics and graphs. I needed the tools of equations and graphs to explain things while he could just talk about economic principles fluidly. Of course I have since moved back to engineering and spend the last 22 years striving to reach the fluid knowledge of electrical engineering that Rich had for economics; including doing a PhD in engineering. I can now talk about electrical engineering as Rich talked about economics, without referring to the tools. This book is what bloomed out of that knowledge.

Contents

Chapter 1

Introduction

Electricity has provided huge benefits for mankind. The small village homes depend on it for lighting and irrigation of crop lands. At the other end of the spectrum, the prime mover of the largest machines in the world have over the past few decades moved away from combustion engines and hydraulics and to electric induction motors. Extremely hot areas of the earth like the Arabian Peninsula have been made habitable with electricity powered air conditioning and extremely cold places like Alaska can attract human populations with electricity powered heating. Thus electricity has become a fine compatriot of humanity. But, if we touch an electricity carrying wire, we will be burnt. Thus it is imperative that as many people as possible have proper knowledge of the limits, dangers of electricity and respect it. This book hopes to disseminate the knowledge of the electric power system to as many people as possible. Calculations are avoided as far as possible. The most useful formulae in electrical power are Ohm's law (1) and it's derivation, the power law (2)

$$V = IR \qquad (1)$$

$$P = VI \qquad (2)$$

Where V = Voltage or EMF (Electromotive Force) in volt units, I= current in ampere units and R = resistance in ohm units and P = power in watt units.

The simple principle of electricity is that generation of power should always equal to the customer demand for it. If this is not balanced, there will be effects on voltage and frequency. The control of this balance is the biggest complication in an AC (alternating current) electrical power system. Nowadays intermittent renewal energy is fast replacing conventional hydrocarbon energy which suddenly made the system even more complicated.

Propagation speed of electricity is affected by insulation. In an unshielded copper conductor, it is about 96% of the speed of light, while in a typical insulated coaxial cable it is about 66% of the speed of light (3 X 10^8 m/s). But actual speed of electrons is near 0 m/s in AC and comparable to putty flow down a wall in DC (direct current). As an analogy to explain this is if a pipe with ends named A and B are filled with table tennis balls. And if the first ball at point A is given a little push, immediately the ball at end B will move. In a similar way, a push of the first slow moving electron at point A of an electrical wire will cause the electron at point B of the wire to move immediately; this immediate action is termed current (I). This action happens at about 96% the speed of light in a bare electrical wire (conductor). If the conductor is covered with in insulation, current flow can drop up to 66% the speed of light.

For current carrying wire, the analogy of the finger which pushed the first ball in the pipe is replaced by a term called voltage or EMF. Thereby voltage is the force which pushed the first electron giving a force to each and every ball all along the pipe. If the finger pushed the first ball with a great force at end A, the tennis ball at point B will jump out and fly quite a distance. The combination of the finger force plus how rough or smooth the pipe is (resistance = R) for all the tennis balls to flow is termed the power (so P=VI, whose units is watts). If the pipe is rough (high R), even a forceful push (high V) of the ball at point A will not result in the ball at point B moving much. But if the pipe is made as smooth as humanly possible (low as possible R) even a slight finger push (small V) will cause the ball at point B to move with close to exactly the finger pushing force exerted at point A.

Another analogy is a powerful BMW next to a tiny Daihatsu on a straight road. The BMW has high power (high EMF or V) and the tiny Daihatsu has low power (low EMF or V). The speed of the car is the current (I); sometimes the tiny Daihatsu can actually beat the powerful BMW if the driver is good enough, that is, the current of the Daihatsu can be greater than the BMW even though it has a low V. When the two cars meet a traffic jam, they are experiencing R.

The 'roughness in the pipe' is termed resistance (R). There are exactly three ways electron flow can be slowed:

1. **Resistance:** roughness of the pipe which restricts the table tennis balls from flowing is equivalent to resistance.
2. **Inductive reactance:** assuming the table tennis balls are replaced with ball bearings and a magnet it placed over the pipe, that will slow electron movement. Note that current carrying wire forming a coil is a magnet by itself; it is called a solenoid. Andre-Marie Ampere discovered the fact that electrons moving in a circle parallel to each other (as in a solenoid) forms a magnet. While in a permanent magnet, unpaired electrons spins line up to each other in regions called domains causes magnetism. This slowing of electron flow (current flow) as it moves into a solenoid is called inductive reactance.
3. **Capacitive reactance:** assuming the negatively charged table tennis balls have built up on the first plate and the second plate is neutral (having an equal number negatively charged tennis balls and stationary positive charged balls). If more and more of negatively charge balls accumulate on the first plate, a critical charge will be reached to enable these balls to have enough energy to jump to the nearby positively charged plate. This accumulation of electrons on one place before jumping onto a nearby plate is called capacitive reactance.

Therefore all three, resistance, inductive reactance or capacitive reactance slows down current flow.

Chapter 2
Electrical Safety

We cannot say current or voltage is more dangerous. The combination of both, the power whose equation is below

$$P = VI \cos \theta \qquad (3)$$

is what causes electric shocks. Where θ or $\cos \theta$ is an indication of the degree to which the current waveform have shifted with respect to the voltage waveform. This variation in the travelling speed of the current waveform with respect to the voltage waveform is call phase shift. The phase shift only occurs in AC moving through a capacitor or inductor and doesn't happen if the AC is moving through a resistor. Phase shift also do not happen in DC.

It takes about 40 volts of force for the negative charge from a current carrying conductor to jump into a human skin. This 40 volts plus or minus will form a sigmoid curve if human resistivity data is collected. People with moist hands will conduct electricity slightly better than those with dry hands. An equipment can have very high power but insufficient voltage for that current to jump into your skin. For example a very big and high powered speaker can have 12V leads but high current (amps) to deliver such huge power to the speakers (as in some dance places). A human can touch this 12V lead even when the speaker is live.

Definition of danger in electrical installation: any source of voltage which is high enough to cause sufficient current flow to the muscles or hair.

Electric currents are always finding a pathway to go to Ground (Earth). If there is an easier pathway for it to reach Ground (Earth) via a copper wire, it will avoid choosing human bodies to reach ground. The ground resistance (earth resistance) must be <100Ω for homes or offices, it must be <10Ω for a steel electric pole and <1Ω for substations and power stations. To put into perspective why electricity loves to go to the ground; if two 10Ω resistors are connected in parallel, the resultant resistance is 5Ω. If four 10Ω resistors are connected in parallel the resistance drops to 2.5 Ω. If eight 10 Ω resistors are connected in parallel the resistance drops to 1.25Ω. Now 1cm^3 of silicon has a resistance of 200kΩ (sand is silicon dioxide SiO_2). If billions of these sand particles are touching a Ground (Earth) rod, they are effectively billions of 200kΩ resistors in parallel and the Ground (Earth) resistance can easily reach less than 100Ω which is the requirement for Ground (Earth) rods of every home, office or factory.

Below is an EXCEL calculation of resistor in parallel following the equation:

$$\frac{1}{R_T} = \frac{1}{R_1} + \frac{1}{R_2} + \frac{1}{R_3} \ldots \ldots \quad (4)$$

1/200,000	1/200,000	1/200,000	1/200,000	1/200,000	1/200,000	1/200,000
0.000005	0.000005	0.000005	0.000005	0.000005	0.000005	0.000005

This was done till column KQ which is adding 303 resistors in parallel of 200,000 Ω each. The final resistance of 303 resistors in parallel is 662Ω. Thus finer grain size means even more 'resistors' in parallel. A highly compacted fine sand around the Ground (Earth) rod will add even more 'resistors' in parallel. A wet ground (mineral filled water conducts electricity but not distilled water) will lower the resistance of each sand particle. In this author's experience one location which is even famous for flooding has a relatively high Ground (Earth) resistance because the water is from a river which has less minerals.

A 240V shock will be more serious than a 120V shock because the force (EMF) for the current to jump into the skin is double. A person dealing with electricity should not be allowed to wear any gold, silver or other conductive ornaments. There are even cases of 6V sending high current through a person's gold ring sometimes even taking the finger off. When this author graduated with an electrical engineering degree and going through the ceremony at South Dakota State University, USA, an oath ceremony had to be gone through where oaths like not designing anything which will endanger humans and down the list is one oath whose statement was, "I will never wear gold on my body."

The weakest point in the human body with regards to ability to withstand electric shock is the heart and brain. So these two portions of the body should always be as far as possible from any electric conductor while performing electrical installation. The CPR (Cardio Pulmonary Resuscitation) is performed on a victim of electric shock. The human body is basically an electric machine so when electricity goes into the body from an external source, the body system is disrupted; electrical signals that instruct the heart to pump are disrupted. The heart 'forgets' how to pump. CPR is done to reteach the heart how to pump. 30 pushes to a point one inch above the bottom of the sternum (meeting point of the ribs), two mouth to mouth blows and then another 30 push. This cycle of 30 plus 2 must be repeated five times before performing a test of the victim's blood circulation. The recommended method to do this test is by placing the pointing and index fingers at the soft region of the victim's neck in-between the harder throat and the hard muscles at the side of the neck. If he still has no circulation, the process must be repeated till the ambulance comes. If there is circulation the victim must be paced in the recovery position; body lying on the side, bottom arm stretched out and other arm on this arm, top leg touching floor and over the bottom leg.

It has to be noted that all metal work used in electrical systems must have a Ground (Earth) wire joined to it. The reason for this is that as mentioned before, electricity loves the ground and if there is a Ground (Earth) wire joined to the metal body, electricity will use it to go to ground rather than a via the human body touching the switchgear (any device used to control electricity flow is called switchgear). The resistance of a human body to electricity flow is too high to be a choice for electricity to use to go to Ground (Earth) if there is a very low resistance (low Ω) pathway like a Ground (Earth) wire. So the human will not experience any shock touching an electricity leaking switchgear which has a ground wire as shown in Fig. 21. One of the main reasons electricity leaks to the body of a switchgear is when plastic insulated wires goes into the switchgear via sharp holes drilled into it's metal body. Cable glands must be used whenever a cable is running through a metal hole.

Chapter 3
History of Electricity

In 1600 William Gilbert coined the word 'electric' which comes from the Greek word 'elektron' meaning amber. He wrote a book named, 'On Magnetism' which was published in 1600.

In 1744, a Dutch scientist named Pieter van Musschenbroek invented the Leyden jar and named it after his town of Leiden (Leyden). This was the first capacitor invented. He was basically trying to capture whatever is electric in a wine bottle. He accidently found that it stored more charge if he held it with his hand. This jar stores static electricity and was used by early researchers to perform experiments in electricity. The jar was a glass bottle coated with a metal foil which doesn't reach the cork top. The jar is partially filled with impure water. It has to be noted that pure water is not conductive. A metal wire passes through the cork top which closes the jar. This wire is connected to an external terminal which is a sphere to avoid losses by corona discharge. Early researchers believed the charge was stored in the water but Benjamin Franklin investigated it and concluded that the charge was stored in the glass container.

In 1752, 46 year Benjamin Franklin (American) with the help of his son flew a kite with a key tied to the bottom of a silk string. They also tied a thin metal wire from the key to a Leyden jar. He then attached a silk ribbon to the key and held it from inside a barn away from the rain. It was a raining day and lightning struck the kite and went down the wet silk string to the key and into the Leyden jar. Ben was unaffected by the negative charge because he was holding a dry silk string (insulator). There was a U shape between the key and the point Ben was holding the string. This is critical in electrical engineering installations; water will follow along the string (or electric cable) from a long way up but will fall at the bottom of the U, thereby the portion of the cable at the next termination or holder of the cable has little water. Thus in all electrical installations, there must be a U just before the next termination point of the wire.

When Ben later moved his hand near the iron key he received a shock because the negative charges from the key went to Ground (Earth) via his body. An arc jumped from the key to his hand. The experiment proved that lightning was static electricity. Ben was lucky to have survived this experiment because a later scientist who replicated his experiment was killed by the lightning. From this experiment Ben came up with the idea of the lightning conductor which later saved his home from a direct lightning strike. One person managed to capture a picture of lightning striking the Eiffel Tower showing lightning choosing to strike only the copper lightning conductor at the highest point and not any of the metal works. The steel metal work is a conductor and is joined to ground via a pile. The pile has a mesh of steel (BRC) and goes deep into wet ground, thereby making it a good Ground (Earth). But lightning chose to strike only the lightning conductor because it can differentiate the resistance difference between the steel framework and the copper conductor which is the lightning rod. Today various shapes of lightning conductor tops have been devised to attract lightning better but the basic principle from Ben is that if a Grounded (Earthed) copper rod is placed at the highest point of a building, it will attract lightning because lightning only wants to go to ground with the lowest resistance pathway it can find. If lightning can find a human body as a pathway, that will be chosen. This happened when three children played football on a rainy day in Kuching, Malaysia which lies on the Equator and therefore has a high lightning rate. It was a big football field but the lightning chose to strike two of the boys who died instantly. The boys have blood in them which contains iron which is a conductor. The lightning can differentiate this resistance difference between four feet of air and a child's body and therefore chose the body as the easier path to enter Ground (Earth).

Ben made a mistake that electricity will move from the positive to the negative and we are still living with his mistake today. The direction of current he postulated is opposite of the actual electron flow which causes current. Books were expensive in those days so when later scientist discovered that Ben actually made a mistake they decided to leave it that way. When electrical calculations are made totally the wrong way round, the answer will still be the same. Ben coined many electrical terms like conductor, condenser (capacitor), battery, charge, positively, negatively, plus and minus.

In 1780 Luigi Galvani (Italian) put a frog leg in-between two different metals. But he believed the electricity came from the frog leg and called it, 'animal electricity.'

In 1800 Alessandro Volta (Italian) realized the frog's moist tissues could be replaced by cardboard soaked in salt water, and the frog's muscular response could be replaced by another form of electrical detection. He disagreed with the notion of Galvani of 'animal electricity' and proved it as he went on to create the Voltaic Pile to produce the first continuous 'current' of electricity or the first battery. Current is a measure of the speed of flow of charge through a conductor. It was first used by Volta to describe the continuous current (like a stream of water) he observed coming out of his voltaic pile which was different from impulses of static charges observed by leaders of science prior to this. The current unit is ampere 'amps' for short. The flow of electric current is observed when a voltage (a force) is put across the conductor. When this same voltage is exerted over an insulator no current results because there are no free electrons to carry the current in an insulator. One ampere is defined as 6.28×10^{18} charges per second. Imagine that many germs or viruses entering the body, which is why one ampere can kill a human being. When current flows in a conductor, heat is produced because of the resistance to the movement (caused especially by the vibrating ions which are atoms with missing electrons) of charge in the conductors. It is for this reason that size of cables needs to be increased to cater for higher current flow.

Georges Leclanché (French) invented a more stable battery called Leclanche cell in 1866 which was widely used for telegraph (invented by Samuel Morse around 1835).

In 1784, Henri Coulomb (French) discovered an inverse relationship between the force between electric charges and the square of their distance.

In 1820, Hans Christian Ørsted (Danish) noticed a compass needle deflected from magnetic north when electric current from a battery was switched on and off. He also coined the Right Hand Thumb rule which states that if the thumb points in the direction of a current, the four fingers indicate the direction of the magnetic field around the wire. The unit for magnetic induction is B in Oerstad.

In 1820, André-Marie Ampère (French) showed that parallel wires carrying currents attract or repel each other, depending on whether currents are in the same (attraction) or in opposite directions (repulsion). This laid the foundation of electrodynamics.

In 1821, Micheal Faraday (English) invented the first electric motor. Electricity was passed through a wire and into another wire that was hooked freely onto this wire. The freely hooked wire goes into liquid mercury (thus free to move) and on to the negative terminal also hanging in the same mercury. A cylindrical permanent magnet (PM) was placed in the center of the mercury bath. Upon switching on this circuit, the freely hooked wire moved continuously around the PM.

After this invention he deduced that if electricity can produce magnetism, magnetism should be able to produce electricity. So shortly after he invented the first motor he also created the first generator. So Faraday basically created the most important inventions of the modern electric world, the motor and generator. He also came up with the Faraday's law of induction states that magnetic flux changing in time creates a proportional electromotive force.

The unit of capacitance is farad, named after him. He coined anode, cathode, electrode, and ion. Faraday constant is the charge on a mole of electrons (about 96,485 coulombs).

Faraday was different from the above scientist in being from a poor family. He was mostly self-educated. At age 14 he started at a book making factory. Here he read all the books in print and became quite knowledgeable. The chief scientist of England at the time heard of his diligence in learning and took him in as a technician. Now he had probably one of the most advanced labs in the world in his hands and utilized his time performing experiments. He was not treated right especially by the wife of the chief scientist. He fought through this and eventually became better than the chief scientist and the queen of England wanted to give him a 'Sir' title but he refused. He is currently generally regarded as the father of electricity.

In 1824, William Sturgeon (English) invented the first electromagnet by winding 18 turns of bare copper wire in a varnished iron horse shoe. His 200 grams electromagnet could lift nine pounds of iron. Because he used uninsulated copper wire the turns around the horse shoe has to be a single layer so he could not increase the number of turns.

In 1830, Joseph Henry (American) improved the electromagnet by using silk thread insulated horse shoe as well has silk thread insulated wires to wind many layers of turns of copper wire around the horse shoe. He built an electromagnet that could carry 2,063 lbs. of weight. Such electromagnets were used in telegraph sounders.

In 1827 Georg Ohm (German) came up with the relationship between Voltage, Current and Resistance: V=IR.

In 1831, Joseph Henry (American) created one of the earliest ancestors of modern DC motor. Unit for electromagnetic inductance is Henry.

In 1833, Wilhelm Eduard Weber (German) together with Gauss developed the first electromagnetic telegraph. Weber in volt/second is the unit for magnetic flux.

In 1834, Moritz Hermann Jacobi (German-speaking Russian) improved on Faraday's vertical motor and Henry's similar horizontal 'rocker' which moved back and forth, to create the first horizontal motor that looked like today's motor. He used battery and his motor to power a boat.

In 1890, John Ambrose Fleming (English) coined the Fleming right hand rule and the Fleming left hand rule. In 1905, he patented the 'Fleming Valve' which is the first diode and this gave birth to modern electronics. Fleming also coined the term Power Factor (PF).

Thomas Edison (American) was the next great inventor. In fact he is the greatest as far as electricity is concerned. What he did is light up the world. Electricity would still been a lab science if not for his combining business, politics and engineering to bring out electricity to everyone. He created the 'Invention Factory' (at Menlo Park) whose function was to make money out of inventions. He gathered many researchers to work as a team to develop products. This was a new concept as previous scientist generally worked alone. Electricity, telephony, motion picture and a host of other things were mostly inefficient lab things till he made them truly useful. Edison was the Steve Jobs of those days. Edison was issued 1093 patents which is the most awarded to any person till today. His favorite was the phonograph or voice recorder, the predecessor to all out recording devices used today. Edison spent 40 years continuously improving it. In the year 2000 authorities of science gathered to pick the most important person for the last 1000 years and Edison was chosen as number one. His recommendation was just hard work, not giving up on the problem and common sense. He said genius is one percent inspiration and ninety-nine person perspiration. He started a company with J.P. Morgan (richest person in USA) and the Vanderbilt (of railroad fame) called the Edison Electric Light Company. This company was started before he invented light bulbs. Today it is called General Electric (GE) one of the top companies of the world and the biggest in the electric industry. The next few in ranking are Siemens, ABB and Schneider. It is important to know these companies because if they are consulted or their parts used, the likelihood for project success will be much higher. His lab was the predecessor to current labs in Google, Apple, Microsoft and Facebook where he stated that there are no rules as long as inventions can be accomplished (way before people had an inkling of this ides). Edison's last patent before he died was artificial rubber. Edison was still experimenting till the last moments of his life.

All the scientists above worked mostly in labs. Edison went to school for only three months starting at age seven. Here he would ask, 'Why?' to his teacher. The teacher got so annoyed one day that he scolded him for interrupting the class. He went home and told his mother this. His mother took him out of the school after a subsequent quarrel with the teacher and decided to teach him herself at home. But Edison cannot be considered uneducated because he was an avid reader. He got this habit from his parents who were avid readers themselves and often read book aloud to their children. Edison's parents bought a science experiment book and he performed all experiments mentioned in it. Thenceforth he was more into experimentation than reading and built a lab in his home for this.

At the age of 12, Edison started work at a train station. He used all the money earned to purchase equipment and chemicals for his lab. He was allowed to move his lab into a train cabin. He also started a newspaper in this cabin. His newspaper reported events of many towns which the train transverses while other newspapers only had single town news. Therefore his newspaper business was very successful. It seems like a simple victory but in those days it was very innovative to realize that trains had suddenly brought in a new opportunity and he took full advantage of it. In today's age it is equivalent to finding a niche in fast expansion in connectivity worldwide.

With this money he made in the newspaper business, he procured chemicals to do scientific experiments since chemicals were the main science of those days. One day there was an explosion and his cabin caught fire. The station manager threw him off the train. One day he jumped onto a train and the station manager caught him by his ear. He said he felt something snapped in his head at that moment and he lost much of his hearing ability. Later he became a rich man and doctors said this problem can be cured but he refused saying the deafness helped him to concentrate on his work. The deafness drove him to reading; he would go into a library and read all the books in them.

One day he saved the station official's son from a collision by an oncoming train; he jumped onto the son to move him away from the tracks. For this the station official rewarded him by teaching him how to operate the telegraph (invented nine years before Edison was born by Samuel Morse).

Edison's next job was being a telegraph operator in Canada. His job was to send hourly signals to Toronto. So he built a device that would do it automatically. This was his first invention. Later he invented an electric vote recorder and displayed it at a Science Fair in New York. People at the NY exhibition were impressed but no body bought his invention. He was frustrated and made a promise that from now on anything he invents must be sellable. Thus he is today known as the 'First Engineer'. A scientist can invest time and money on abstract science but an engineer must make money thereby and engineer is half scientist and half businessman. Such a philosophy of Edison is what drives all high tech company today. Engineers in those companies must project earnings for every research or investment they do.

Edison strung a grid of underground wires in New York, making it the first grid in the world to supply electricity to his Edison bulbs in homes. Edison and his team tried out about 6000 materials before settling on carbonized bamboo as the filament in Edison bulbs. Edison would spend hours and days in his laboratory concentrating on some experiment or problem; milk, bread or tea was pushed into the room from under the closed door, but they were untouched until he solved the riddle that was in his mind. So great is the riddle that science demands.

Edison's business became so big that it spread all over USA and moved into Europe; this was rare in those days when it was normally European inventions that moved to USA. Satellites have recently taken numerous pictures of the earth at night and combined those images without cloud cover. The whole earth can be seen to be lit up, of course the advanced countries and cities being lit up more. And all this is due to Edison bringing electrical technology out of labs and turning it into a business.

In Edison's office in France a brilliant man named Nikola Tesla joined his company. The manager in France was so impressed by Tesla that he wrote a letter to Edison stating that he knows only two great men Edison being one and this man named Nicola Tesla being the other. Tesla was asked to move to Edison's headquarters in New York. The first task Edison gave him was to improve the gensets to a particular specification. As great an inventor as Edison was, he tried to achieve those specs but failed. So Edison was so confident that Tesla cannot achieve those specs, that he actually promised him $50,000 ($500,000,000 in today's money) if he can achieve it. Tesla achieved it in a short while and went to Edison to collect the money. But Edison said it was just an American joke. The frustrated Tesla resigned. He found employment digging drains in New York. While doing this, another rich man named Westinghouse found him in working the drains and offered him a lab and eventually $1,000,000 for most of his patents, including the AC system and the induction motor patent. The induction motor was invented by Tesla much earlier on. In those days carbon brushes of DC motors was of low quality but today a carbon brush in a DeWALT or Hitachi drill can last more than 20 years. So the commutator system consisting of a pair of carbon brushes and copper electrodes separated by air gaps was the biggest worry for industrialist. A motor suddenly not running in a factory can cause lots of damage to products. Therefore Westinghouse was confident Tesla's brushless induction motor will be easily sold. From the Westinghouse lab, most of the AC system we use today was invented by Tesla. Tesla lost a huge financial windfall because he tore up the contract which states that he would get a certain amount of money for every kW of electricity sold which would have made him one of the richest person in the world.

Of course Edison was not at all happy with the brewing AC system. DC was a business friendly solution and actually a stable one. DC cannot travel far because it cannot be stepped up with a transformer.

How AC and transformer helped in enabling long distance electric transmission is as follows. Say in a hydroelectric power station, the power (P) is from the falling water. This power has two components voltage (V) and current (I):

$$P=VI \qquad (5)$$

A hydroelectric generator produces electric power normally at 11kV. A transformer is used to push up the V to a very high level, like 500KV. Looking at equation (5), this will cause the I to go down because P cannot change since it is the power of the falling water. The electrical energy then goes to the overhead lines where energy loss in transmission is defined as:

$$P_{loss} = I^2 R \qquad (6)$$

Therefore as power is sent to a faraway city say 1000 miles away; putting the low I into equation (6) will give a low power loss. Without this being taken cared of, the overhead lines will be red hot and by the time it reaches the city, there will be very little power left. Actually equation (5) is the same as P=VI but it is an equation generally used by electricians to calculate power loss because it emphasizes the importance of current and not voltage or power that plays the major role in power loss. In the early days everyone who dwelt with electricity, from technicians to scientists were termed electricians. Power loss in a single phase is given by equation (6) but in three phase the equation is as follows. In Fig. 1 below we assume Y connection but using Δ connection as in Fig. 2, the final equation is the same.

$$P = \sqrt{3}V_L I_L \qquad (7)$$

$$But\, V_L = \sqrt{3}V_P \qquad (8)$$

$$P = \sqrt{3}\left(\sqrt{3}V_P\right)I_L \qquad (9)$$

$$So\, P = 3V_P I_L \qquad (10)$$

$$But\, V_P = I_P R \qquad (11)$$

$$So\, P = 3\left(I_P R\right)I_L \qquad (12)$$

$$But\, I_P = I_L \, using\, Y\, connection$$

$$P = 3\left(I_L R\right)I_L \qquad (13)$$

$$P = 3I_L^2 R \qquad (14)$$

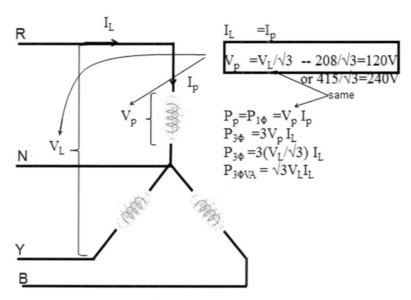

Fig. 1: Y connection

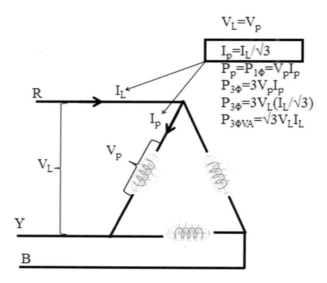

Fig 2: Delta connection

But Edison's business model was good; a DC genset could supply electricity only around a one mile radius. He could sell electricity and gensets. Using AC a single genset can power the whole New York state. However DC has one advantage in that it is very stable. If a DC line is running across a city, anyone can tap from it or send power into it as long as their generator is producing the same voltage. AC is comparatively very 'nervous'; voltage as well as frequency must be the same before joining a genset to an overhead line. That is, a genset need to be synchronized to the overhead line current before joining. Also lots of factors can cause phase shifts of voltage and current in AC (voltage and current not travelling at the same moment) and these causes problems for which reactors (big inductors), capacitors banks and playing around with excitation of synchronous motor are required to solve. So Edison was right in stating about AC, that it is a useless complication; but the words he used were too extreme. It is definitely an increase in complexity but is useful; the cost reduction in the electrical system and therefore the final bill people paid was too good with AC.

Edison used lots of resources to prevent the advent of AC, just as any businessmen would to prevent something which will disrupt their business. He put up newspaper articles stating that AC is dangerous and even built the Electric Chair to execute people using AC; just to show how dangerous AC is. But of course a DC powered electric chair would do the same thing.

But Tesla got his break in the Chicago World's Fair. The US government wanted to celebrate the 400th anniversary of Columbus's voyage to America by having a World Fair at Chicago. Quotations were given by General Electric (Edison and JP Morgan's company) as well as Westinghouse. But Westinghouse's quotation, using AC was about one quarter the price so Westinghouse was chosen. Edison who got disappointed stated that his Edison bulb cannot be used for that project so Tesla invented a double stopper bulb (manufactured for only one year because the patent on Edison bulb expired the next year) and also displayed the early form of the florescent tube for the project.

It was at the Chicago World's Fair that humans from all over the world saw for the first time so much brightness in the night sky and that shot Tesla to fame. JP Morgan thenceforth decided to get General Electric to use AC with Tesla's help. JP Morgan's argument is that people in Germany were using Tesla's system to transmit electricity over 160 miles which was not yet achieved in the USA despite as Morgan said, having Tesla right below their noses. Edison having lost the 'war of currents' went away from the electric industry and focused on inventing the whole movie industry.

Tesla was later free and was able to move to Colorado to solve the mystery in his mind. While working for Edison, he observed purple light emanating from electric overhead lines for a short moment after the big DC switch was turned on. After a while the purple light would vanish. But if a man stood below the overhead line while the purple light was observed, it would travel down through the man to ground, killing him instantly. He can estimate that this purple light was not caused by electrons because it seems to travel much faster than electrons. As mentioned earlier in this book, electron flow in DC is as slow as putty flow. Tesla used all his resources to discover what this purple light is and finally concluded it was energy emanating from ether (empty space). He called it 'radiant energy.' People always thought empty space is empty but Tesla stated that it is full of energy. This purple light emanates when a powerful DC is switched on. So he created a Tesla coil using capacitors and inductors and a spark gap to achieve a continuous switching of a powerful DC current (replacing the big DC switch used at Edison's company) to achieve free energy.

Tesla then moved back to New York because JP Morgan heard about his achievements in Colorado and invited him to build his research center in New York. JP Morgon financed the big Tesla coil named Wardenclyffe Tower for $150,000. JP Morgan was only interested in having a way to communicate with Europe. Tesla have already invented the long distance radio transmission (in Europe Marconi is named the inventor but the US government have officially named Tesla as the inventor; Marconi used 17 of Tesla's patents to recreate it). Tesla told JP Morgan his tower can achieve telecommunication with Europe mostly with through-earth communications. Before fully completing the tower, Marconi's radio invention came out and was basic and thus much cheaper. Also at this time, JP Morgan went for a site visit to the Wardenclyffe Tower where Tesla informed him that his tower can also create free energy which can be broadcasted to the ionosphere. Ionosphere is a layer of ions (atoms with more or less electrons) surrounding the earth; note all ions even insulator ions are conductors. His plan was to send electricity to the conductive ionosphere where it will flow continuously. This electricity can then be tapped from the ionosphere for example by a home in Europe or a ship off Japan by utilizing a small resonating tesla coil. JP Morgan asked how his company would charge (meter) the home in Europe or the ship off Japan for their electricity. Tesla said it would be free and this got JP Morgan was very angry; businessmen do not want to give away things free. He pulled out all funding. Actually what Tesla was building was to enable not just electrical power and communication, he also stated that images or movies can be transmitted in real time across continents. Thus Tesla was about to bring the world to beyond year 2015 technology in the 1920s.

Tesla was dejected but build a car powered by this free energy. There was no battery and no combustion engine and it could run. Others who saw the car were amazed and asked if the car worked on 'Black Magic.' So now the car is called, 'Black Magic car'. Tesla later wrote how he made this free energy in a paper and cut it into a jig saw puzzle. He put a piece of the jigsaw puzzle plus a cover letter and sent it to all the major countries of the day. He was frustrated by the numerous wars among countries so he wrote in the cover letter that if you all become friends you will have my energy.

Tesla also invented other things like free fertilizer. He said 78% of air is nitrogen, and fertilizer is just a way to get nitrogen into plants so he questioned why can't nitrogen be tapped from the air and used as fertilizer. He thus invented a free energy machine into which farmers have to hoe in soil and out comes nitrogen filled soil.

Tesla's induction motor can be made to last for a very long time with almost no maintenance. We can observe in industry today that the good quality ones almost never fail. This author have seen a 25 HP (1 HP or horsepower = 746 Watts) motor (made by Baldor) carrying 500 kg running 24 hours a day, every day for 14 years; getting about an hour break each day because humans cannot load the machine fast enough. And the process requires pouring heavy rain amount of water and the motor is at the bottom-most point of that machine. This was at the Western Digital hard disk factory. But it must be noted this author has also seen induction motors which look exactly like the Baldor motor emitting varnish gasses upon first energizing with 415V. Therefore not all induction motors are the same, brand makes a whole lot of difference. It is mostly the difference in the quality of the varnish around the coils and bearings which makes the difference. The quality of a good engineer is to keep up (coffee break talks etc.) with which company produces quality parts for their projects.

Imagine a Tesla world, one can drive as far as one can and will never need to refuel. The engine is an induction motor which almost never fails. If car driver stops for a meal, it will be quite cheap because fertilizer is free. There will be no need for the electric grid wires because all electric transmission will be wireless. It will be an utopian world with no greenhouse gasses; Mother nature will love humans. But the huge oil and gas industry will collapse, electric utilities will collapse, the whole fertilizer business will collapse and companies making overhead and underground cables will collapse. This will give a hard time for the rich people of the world. If Tesla's systems were adopted, no one will be too rich and it will take very little money for the poorest to survive. So everyone will be about equal in wealth. Of course capitalist will not agree to implement Tesla's innovations; they want to keep their status quo of being above the rest in wealth and power. So Tesla's name was marred as far as possible and even taken out of science books for a time.

Today researchers all over the world today are peering into his patents hoping to replicate his wonders. In the 1920s he achieved frequencies which the latest telecommunication companies cannot achieve today. After the decommissioning of the Wardenclyffe Tower by JP Morgan, Tesla himself stated, his inventions were ahead of his time but one day humans will use his inventions, humans needed to see the consequences of inefficient and earth polluting methods before finally adopting his methods. As far as the usage of his induction motors which run our home fans, air conditioners, factory motors all the way to ship propellers and the largest cranes in the world; his prediction has already come true. The AC system he invented is also currently the most expensive installation of mankind today (worth around six trillion dollars). Therefore it is a safe prediction that all his inventions will one day be used by all of humanity.

Al Gore (former vice president of USA) made a documentary in 2006 called, 'Inconvenient Truth' where he stated that if a bore is made in the ice of the coldest portion of the Antarctic (where the ice never melts), CO_2 level data for 650,000 years can be measured; there is a visible layer indicating each year of snow. The chart showed a CO_2 level that was mostly flat till the 1900s, and then it shot up far higher than it has ever been in 650,000 years (due to industrialization and combustion engine vehicles). It is possible that God sent Tesla to solve the problems of mankind at the right moment in history but the rich businessmen were not willing to forgo their wealth and power to achieve a pollution free world. An IEEE magazine reporter asked Albert Einstein how it felt to be the smartest person alive. Einstein replied that he does not know and stated that question have to be asked to Nikola Tesla. Another interesting thing about Tesla was echoed by Steve Wozniak who said all the best things he did came from not having money and not having done it before. Some companies have huge research budgets to develop things but it often takes a 'Tesla' to get things done with often a miniscule of that budget.

Chapter 4
DC and AC currents

DC current can be equated to a pipe carrying water. The pump is equivalent to generator. If the pipe is plastic and is pressed, the water within will have a higher pressure; this is equivalent to a resistor (or capacitor or inductor).

AC current is actually moving back and forth. So theoretically an electron experiencing AC voltage is actually moving back and forth at a frequency set by the power supply utility. An electron therefore stays at one point of an AC wire forever. This back and forth movement is utilized to run a motor, a bulb, a heater or any other load.

In a three phase induction motor, three phase wires sequentially magnetizes the six coils (in most case since most induction motors are four pole motors) located in the stator of the motor resulting in the turning of the squirrel cage rotor. In a four pole motor each phase is joined to two solenoids therefore it has four poles (N,S,N,S). Tesla demonstrated this sequential magnetization of solenoids in a circle to get a rotor turning in his famous Columbus Egg demonstration at the Chicago World Fair. The egg was made of copper and it rotated as the solenoids were energized (magnetized) sequentially around it.

Chapter 5
Generator Principle

The generator works on the basic Fleming Right Hand rule. This rule states that the thumb the forefinger and the middle finger represents the F (Force), B (Magnetic field from magnetic North to South) and I (the direction of the current in the conductor). Remember this by holding the right hand as a gun and the three fingers are FBI (American FBI use guns) top to down. This is shown in Fig. 3. Fig. 4 is the opposite of this the motor principle. When a magnet passes a conductor, a current is induced, following Fleming's Right Hand Rule as shown in Fig. 3. Consequently when current is flowing in a wire in-between a N and a S magnet, there will be movement of the wire, this is the motor principle. In electrical engineering there are many two rules like this it, it is easier to remember one and know that the other is the opposite. This author for example just remembers that left is for motor.

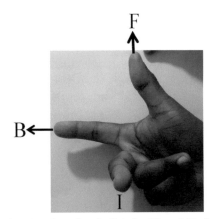

Fig. 3: Fleming's Right Hand Rule

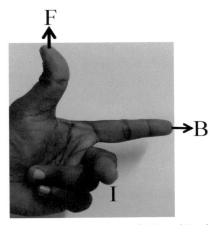

Fig. 4: Fleming's Left Hand Rule

If more current need to be generated, one way would be to have a long wire and a long magnet passing it. But this is not practical so the long wire is formed into a coil and a short magnet and powerful magnet is passed across it to produce higher current generation. In a standard generator today three wires goes into it because if a single wire goes into it, only one AC waveform can be generated and the space within the generator is not fully utilized. Some have tried even more than three wires going into the generator but it was later found out that Tesla's original design of three is the optimized solution.

Theoretically, four or five phases would be better utilize the space within the stator of the generator, but most motors in use currently are made for three phases and these will not run on four or five phase power. Other characteristics of a three phase system are:

• The phase currents will cancel out one another and will sum to zero in the case of a linear balanced load (like an induction motor) where all phases uses the same power but being 120 degrees apart they cancel out each other resulting in zero voltage and current in the return path or neutral.

This summing makes it possible to eliminate or reduce the size of the neutral conductor. That is, if L1,L2,L3 (R,Y,B) carry the same current and voltage the sum is zero and therefore at the N (neutral) wire V=0V and I=0A. Even if the load is not an induction motor and therefore not balanced, the N current is reduced. For example at one particular moment if L1(R)=6A, L2(Y)=-4A, L3(B)=3A. The sum of the three phase current is 6+(-4)+3=+5A, so the neutral current at that moment is N=-5A (the minus sign indicates that the current is returning to substation).

Another case where people are in a bit of confusion regarding the above is in the calibration of MSB (Main Switch Board). The purpose of a MSB is protection. Below 100A, a three phase Distribution Box (DB) is used for protection (as in a home) above 100A a MSB is used. For a country or region, a Substation is used; note Substation for protection as opposed to Power station for power generation. Going from DB, MSB to Substation the equipment gets more and more complicated. In substation the latest computer and fiber optics technology is utilized. Fig. 5 is the layout of the protection CT (Current Transformers – to measure current) in a MSB. And Fig. 6 is the Protection Scheme (like a protocol). To calibrate the MSB first the L and N leads of the Earth Fault (E/F) Relay is disconnected. Then 100%, 150% and 200% of power company approved load is injected into the bars to test if the MSB will trip as planned. The O/C relay will not give the trip signal to the ACB (Air Circuit Breaker) which is the main cut-off switch for a building, at 100% injection; it only trips a at 105% loading. If it trips at 100% current, the O/C relay is rejected. At 150% loading, it will trip at a specified number of miliseconds, at 200% loading, it will trip at another specified number of miliseconds. If the tripping time is out by a little, the calibration personnel can use a tuner in the O/C to bring it back to the specified timing. If it is out by a lot, the calibrator need to reject the O/C relay. This all seem logical. We switch off the power company's incoming and use current generated ourselves to trip the ACB and detect if the tripping timing is correct.

But for the E/F relay, we inject the same points of the current carrying bar with 10% injection and the E/F will send a trip signal to the ACB. Many ask why is it tripping at 10% while normal consumption current is 100% (or at least 70-80%). Fig. 6 is the full setup for performing calibration. In the partial daigram in Fig. 5, all three phase wires are looped with the N wire. The three phase has current going out of the CT and into the OCs then out of the OCs. The N wire has current going the other direction. So the three live wires' current added together is equal to the N current in the opposite direction. That leaves zero current at the wire going out of the four wires to the E/F relay in normal cases. But if there is an earth fault, like a live wire touching the body of a refrigerator connected to L1 (R) phase, some of the current from L1 (R) phase has leaked to Ground (Earth). Now the N wire has less than the combined incoming of L1, L2 and L3 (R,Y,B). Note the combined L1, L2, L3 (R,Y,B) means adding at that moment of time (of the AC sine wave) probably a positive current in L1 (R), a negative current in L2 (Y) and a positive current in L3 (B). So E/F relay will send a signal to trip the ACB if 10% current of approved (by power company) company flowing through the wire coming out of the four wire joint.

Note the current coming out of the CT is much reduced from the actual current but is exactly propotional to the large current flowing in the current carrying bars. In the last few paragraphs it is mentioned as if the large current is going through the OC and E/F relay but it is actually the propotionately reduced current that is doing the work. Electronics cannot work with large currents; they will get fried. The protection system is operated by the propotionate low current but the actual large current causes the triggering and tripping. In all power systems CT are used to propotionately reduce current and VT (Voltage Transformers) are used to propotionately reduce the voltage so electronics and computer technology can determine what actions or reaction to take

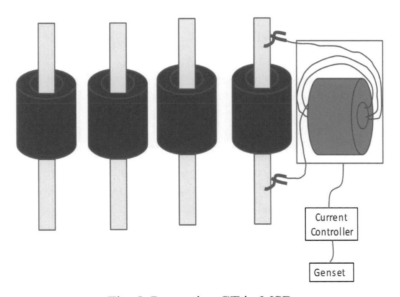

Fig. 5: Protection CT in MSB

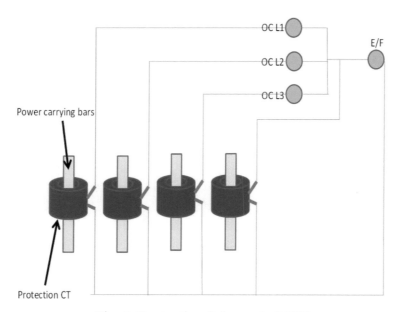

Fig. 6: Protection Scheme in MSB

The voltage at the N is zero because voltage in each phase is controlled by the power company to be fixed to 120V (240V British). In fact that is the job of this author while he worked in a power utility to maintain the main grid voltage to be 275 kV. This was done by continuously starting more generators or energizing or de-energizing reactors which are big coils located at the Transmission network substation (note coils=inductors=reactors).

If the highest voltage is constant all the voltages down the line will be constant. Voltage is also controlled at LV (Low Voltage the opposite is HV or High Voltage) substations which are considered the distribution network, mostly by using capacitors and inductors. So voltage at the consumer (home, office or factory) is fixed at 120V (240V British) per phase so at any moment of time the sum of the voltages is 0V. This can be observed in the Fig. 7; taking voltages to be full (top of sine wave) or half (half way down the sine wave). At 5ms L3(B)=+1, L1(R)=-1/2, L2(Y)=-1/2. So at 5ms, $V = +1 - \frac{1}{2} - \frac{1}{2} = 0V$. At 8 ms, L3(B)=+1/2, R=+1/2,

Y=-1. So at 8ms, $V = +\frac{1}{2} + \frac{1}{2} - 1 = 0V$. Similarly at 15ms, L1(R)=+1/2, L2(Y)=+1/2, L3(B)=-1/2. So at 15 ms, $V = \frac{1}{2} + \frac{1}{2} - 1 = 0V$. If an oscilloscope is used at every moment of time, the V will add to zero.

For current in a in a factory where the only loads are induction motors, the current load on all three phases are always the same, therefore the N wire in that factory will record zero current and zero voltage. In other installations where current is not perfectly balanced among the three phases, the remainder will flow in the N wire.

By looking at the waves in Fig. 7, it is impossible for the sum of current in L1, L2, L3 (R,Y,B) to sum up to a value beyond the peak current in any of the L1,L2,L3 (R,Y,B) phases. Of course Fig. 7 is the voltage waveform. The current waveform is in most cases controlled by the power company to be in phase (crosses the X-axis at the same times) with the voltage waveform but the height of the wave (amplitude) of the wave of each phase is not fixed as in the voltage waveform because voltage is controlled by power company. The current waveform is controlled by the customer. Say a customers in a single phase home connected to L1 (R) phase may have lots of high electricity usage equipment and another other single phase home connected to L2(Y) phase home may have very little electricity consumption. This will result in the current waveform for L1(R) phase to have a high amplitude and the waveform for L2(Y) to have a low amplitude. In later part of this book there will be details of a new phenomenon called harmonics in electrical systems caused by many modern equipment which changes the fact that N current cannot be higher than the current in any of the three phase cables.

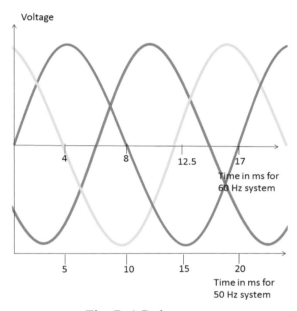

Fig. 7: AC sine waves

But in most cases the N need not be three times the size as is required for a pipe carrying water from three pipes.

- Fig. 7 shows the voltage graph. The current graph will be in phase with the voltage graph if the power goes through a resistive load. The voltage and current graphs experience a phase shift only if it travels through a capacitor (current graph faster) or inductor (current graph slower).
- Power to an induction motor uses the same amperes in L1,L2,L3 (R, Y, B) phases so the motor vibration is less. The vibration principle can also be applied to three phase generator and wind turbines which tend to have three fins.
- Three phase motors are simple because they rotate synchronously with the magnet passing the R, Y B coils in a far-away generator (or synchronously with a small slip as in induction motor).
- Three is the lowest number to exhibit all the above properties.

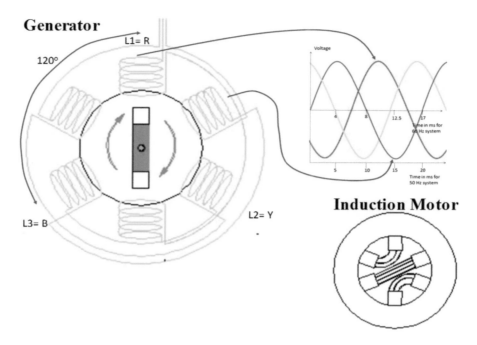

Fig. 8: Three phase generator and induction motor on right

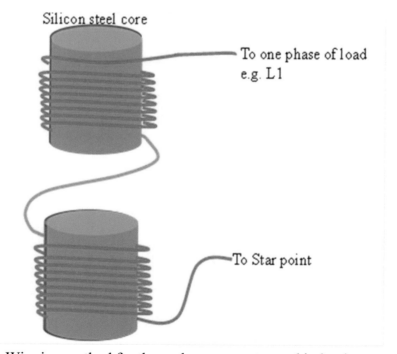

Fig. 9: Winging method for three phase generator and induction motor

The three wires are named L1, L2, L3 (R, Y, B) forms six coils; each wire forms a clockwise coil as well as an anticlockwise coil as shown in Fig. 8. Electron flow (opposite of current) into a clockwise coil produces a magnetic N (North) and electron flow into an anticlockwise coil produces a magnetic S (South). One end of all three wires is tied in a knot called a star point. Note as mentioned previously there are three ways to achieve zero volts:

1. Joining all three phases together
2. Neutral
3. Ground (Earth)

This star point is then connected to a transformer before going to Ground (Earth); thereby providing a high impedance Ground (Earth) or elevated N Ground (Earth), for purpose of protection. The other end of the L1, L2, L3 (R, Y, B) wires supplies power to homes and industries. This is depicted in Fig. 11. The output of big power generators are generally at 11kV to 13KV.

Three Φ motor with eight pole per Φ
The N is connected as a Star point in all big generators

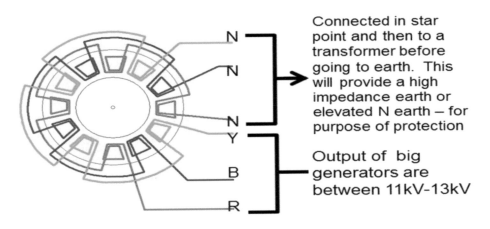

Connected in star point and then to a transformer before going to earth. This will provide a high impedance earth or elevated N earth – for purpose of protection

Output of big generators are between 11kV-13kV

There is a CT on each of the six wires. Terminals of these CTs are connected to protection system

Fig. 10: Wiring in a generator

To remember the polarity formed by the coil, it is looking at electron flow (opposite of current flow), a clockwise turn will form a N and an anticlockwise turn will form a S. Say only one wire goes into the stator (the stationary outside) of the generator, it forms a clockwise coil on the top and an anticlockwise coil at the bottom. As the rotor (the rotating inside) magnet passes the top coil, electricity is generated as a sigmoid curve. Initially a low voltage as the magnet approaches the clockwise stator coil, the electricity gradually increases till it reaches peak voltage V_p as the magnet is fully under the coil. As the magnet moves away from the coil, the voltage decreases till it reaches zero volts.

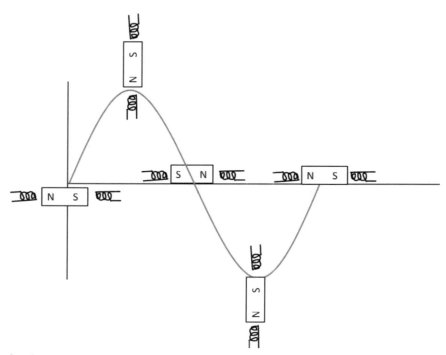

Fig. 11: Waveform of voltage generated as the rotor magnet passes one complete cycle around a generator

As shown in Fig. 8, when the magnetic N of the magnet passes the clockwise coil of L1 (R), a positive high is generated on the L1 (R) phase wire. The magnet then passes the L3(B) anticlockwise coil to generate negative high for L3 (B) phase wire. When the magnet passes the L2(Y) clockwise coil a positive high is generated in the L2 (Y) wire. Next it passes the anticlockwise L1 (R) coil generating a negative high in the L1 (R) wire, consequently a positive high in the L3 (B) wire and a negative high in the L2 (Y) wire. And goes back to produce a positive high in the L1 (R) wire. Thus as the magnetic rotor physically turns a complete cycle one complete cycle of AC waveform is generated.

Because the clockwise coil of L1 (R) phase and the next clockwise coil of L2 (Y) phase are 120 degrees apart, the phase to phase angle of the waveform is also 120 degrees apart. For example the peak positive of L1 (R) to peak positive of L2 (Y) is 120 degrees apart. In a 50 hertz system used in most countries, the magnet turns 50 complete turns in one second. In USA and Japan the generators turn at 60 cycles per second.

A permanent magnet cannot be used in large generators because lots of heat is created inside the generator and permanent magnets will lose its magnetism in the presence of too much heat. Thus an electromagnet is used as the rotor, whose electric power is provided via carbon brush. The slight increase or decrease in the DC current value sent to the rotor coil is used by power companies to control the overall grid frequencies. More magnetism (resulted by sending more DC current in the rotor) in the rotor coils will make it slightly harder to turn and less magnetism (less DC current) will make it slightly easier to turn, which results in phase shifts in the outgoing waves. Larger loads connected to the generator will also slow down the rotor because as more current is drawn from the stator coils, making them more powerful solenoids whose magnetic flux acts like strong strings from these stator coils to the rotor coils or squirrel cage bars thereby slowing down the rotor.

One complete cycle at 60 hertz takes 16.6666 milliseconds ≈17 ms (20ms British for 50 Hz) which is the period of the waveform generated. At low voltage, the single phase peak voltage is controlled to be 169V (339V British). The RMS voltage or 120V (240V British) value is got by placing a voltmeter between any phase wire L1, L2, L3 (R, Y, B) and N or G (E) is derived from this by multiplying the peak voltage by 0.707 (this number can be remembered by the existing the Boeing 707). Voltmeters can only detect the peak value. In analog voltmeters the reading needle is simply bent from 169V to 120V (339V to 240V British) and in digital voltmeters a simple multiplication by 0.707 is done to give the RMS voltage. The phase to phase voltage (placing voltmeter probe on any two of the three phases wires L1, L2, L3 (R, Y, B) is 208V (415V). The phase to phase voltage can also be mathematically derived by multiplying the phase voltage 120V (240V British) by the square root of three, that is $120\sqrt{3} = 208V$ ($240\sqrt{3} = 415V$ British).

The three wires coming out of a big generator at 11KV (phase to phase voltage by convention) is stepped up to a very high value like 275KV or 500KV. The current is therefore reduced. The reason for this is as follows. The power of the falling water in a hydroelectric power station has two components, V and I following the formula:

$$P = VI \qquad (15)$$

When V is increased with a transformer, I has to go down because the P is the water power which is not modified by the transformer. And when this low current value is placed in the power loss equation:

$$P_{loss} = I^2 R \qquad (16)$$

results in a low power loss over the transmission lines, which is wired all the way to the population centers. Here in the cities, it is stepped down to usable voltages. A typical high voltage (HV) tower has four wires carrying current per phase. This is called bundled cables, normally two wires bundled but it can be up to four wires bundled. Thus three phases requires twelve wires.

The current in each of these 12 wires can range from 70A to 110A but it can carry the huge power used by the whole grid. This is because the voltage is very high up to 500kV or 1000kV. 70-100A is actually quite low compared to a car battery which sends out 70 amps upon starting. But the car battery is 12V which is safe for humans to touch even when the car is starting up because it takes about 40V for current to enter a human hand or body.

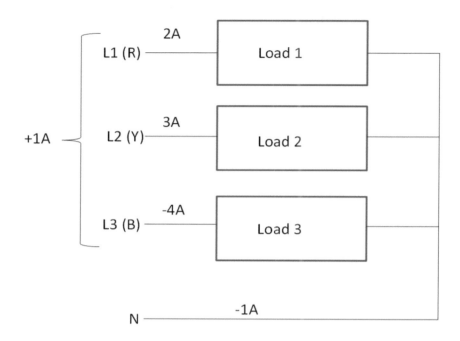

Fig. 12: Housing three phase wiring

As in Fig. 12, the power coming out of a generator is sent to homes where the L1, L2, L3 (R, Y, B) is connected to three homes. That is one phase for each home; this is an example, in actual case, sometimes two, three or four homes can share a single phase depending on the incoming wire size. Coming out of each home is N and all three N wires are joined such that there is only one N wire goes back to the Low Voltage (LV) substation transformer star point and G (E). If the N is broken in-between L2 and L3 phase (Y and B phase), L1 (R) phase and L2 (Y) phase will be shorted providing a voltage of $120\sqrt{3} = 208V$ ($240\sqrt{3}$ = 415V British) for the first two homes. The N wire in these two homes has become Live. This will cause equipment in L1 (R) and Y (L2) phase homes to be damaged because they will be getting twice their rated voltage. Actually it will not be exactly double, initially; each home will have parallel circuit (all homes and all installations are connected in parallel) but the two parallel connected homes will now be connected in series. So 415V supply will drop over each 'resistor' (home). If the two homes have exactly equal load, each home will get half the voltage and nothing burns. But it one home has a much higher load than the other, in a split moment of time, the high load home will get a high voltage, blowing up all the equipment in this home. Now there will be no more load in this home so it is just 415V powering the low load home, destroying all the equipment in this home also.

If N is broken in-between L1 and L2 phase (R and Y phase), the L1 (R) phase home will not get any power since there is no return path to the substation transformer N and G (E). L2 and L3 (Y and B) homes will not know something happened as everything will be running as normal. L1 home people will then call the power company and nothing gets blown.

If N is broken in-between L3 (B) home and outgoing N, the N join of L1, L2, L3 (R, Y, B) homes are now connected as a Star point on the N side which is one of the three ways to achieve 0V. Note there are three ways to achieve 0V, namely, N, E and Star Point. So when a Star point is forced upon the three homes it looks alright because in three phase induction motors, the outgoing wires are sometimes connected in a star point. But there is a problem in that in an induction motor the three coils use the same amperes while the three homes do not use the same amperes. As in Fig. 12, in the incoming current is (+2A) + (+3A) + (-4A) = -1A. At the star point the current will be +1A so on the L3 phase there will be a (-4A) – (+1A) = +3A across the L3 home which seems alright but if the current is -9A at star point and +9A at L3 home, the current across L3 home will be (-9A) – (+9A) = -18A which is higher than the 9A that home normally takes in so the equipment in that home will all burn. Power company staff reported to this author; that if an outgoing N wire broken by a lorry hitting it, will cause the home appliance to be burnt. Initially this author said it cannot be because the outgoing side of the three homes will be a star point like some induction motors. But as more of the power company staff reported the same thing this author realized it must be the current differential in the three homes that cause the appliances burning while in an induction motor the three coils use the same amps.

It is for this reason that power companies are always concerned about N being broken. A power company's Linesman's main job is to ensure the N is not broken anywhere in the system. One university engineer informed that a few whole labs of computers blew up due to a faulty N connection.

Synchronizing generators

When two generators need to be joined they must be synchronized. The schematic for synchronizing is shown below in Fig. 13.

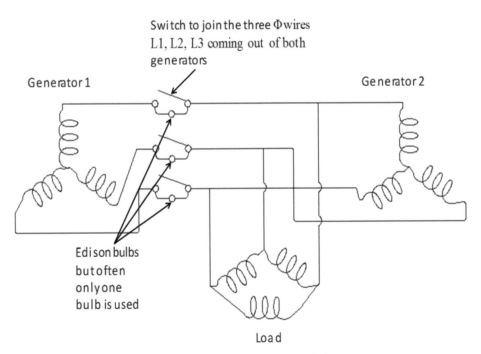

Fig. 13: Generator synchronizing

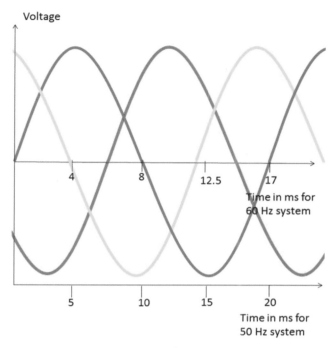

Fig. 14: AC sine waves

What happens here is that when say L1 (R) phase wire of generator 1 and L1 (R) phase wire of generator 2 are joined to the two terminals of an incandescent bulb (Edison bulb), it will light up if the two generators' waveform is not in phase, that is, just like connecting the incandescent bulb to L1 (R) and L2 (Y) phase of the first generator where the voltage will be $120\sqrt{3}=208V$ ($240\sqrt{3}=415V$). As can be seen from Fig. 14 at most moment of time (the X-axis) there is an amplitude difference between L1 (R) and L2 (Y) phase. This amplitude difference is energy to light up the bulb. But if the DC going into the solenoid rotor of generator 1 is played with (increased or decreased) the rotor will at some moment rotate at the same speed as generator 2 because the variation of current in the rotor will make the rotor more or less magnetic. When the rotor is more magnetic, it will turn slower across the stator coils (which have recently become electromagnets also because it is a coil with generated current flowing in it) and when it is less magnetic it will turn faster across the stator coils. The three wires of the stator carry the generated currents out to the loads. For example if the DC is increased, the rotor's magnetic field will interact with the stator's magnetic field to slow down the rotor thereby slowing down the AC waveform produced. So the rotor can be slowed till the waveform produced reaches the speed of generator 2. At this moment the light bulb will go dark because the waveforms from the two generators will be in phase and L1 (R) wire of generator 1 and L1 (R) wire of generator 2 can be joined. Normally only one bulb is joined instead of three to confirm all three phases are synchronized. This is because in a rotating generator if one phase is synchronized, the other two phases has to be synchronized.

Chapter 6
Understanding 1Φ and 3Φ power

If the ac is 120V (240V British), this is the RMS (root mean square) value which will give an equivalent power as a dc with 120V (240V). The reason for inventing this RMS value was to make equivalence between the supplying DC to loads and the newly introduced AC to loads. A 120V (240V British) lamp can run as safely with 120V (240V British) DC or 120V (240V British) AC. The actual measurement of equivalence can be made with an electric heater supplied with DC and another one supplied with AC. Say of 120V (240V) AC required 10 minutes to boil a bucket of water, the AC required to perform the same function will require $V_p = 170V$ ($V_p = 339V$). Of course technicians will not be happy to switch over to AC which had a different number. So V_{RMS} was created to as $V_{RMS} = V_p \times 0.707 = 170 \times 0.707 = 120V$ ($V_{RMS} = V_p \times 0.707 = 339 \times 0.707 = 240V$). The actual voltage of $V_{RMS} = 120V$ ($V_{RMS} = 240V$ British) varies between about -170 volts and +170 volts at 60 Hz and -339 volts and +339 volts at 50 Hz (British).

Why is 50 Hz used by most of the world? AC is actually switching on and off and at 10 Hz rotation of the generator, an incandescent bulb can be clearly seen to switch on and off. As the generator is made to spin at increasing frequencies, it was noted that the human eye cannot perceive the flickering above 55 Hz so Tesla decided to use 60 Hz in USA. But when this technology was exported to Europe, the Europeans did not want to follow a convention from the USA, being supposedly more civilized, plus the fact that they preferred to be more metric. Since 50 is 100 divided by 2, they chose 50Hz. Europeans colonized most of the world so most of the world is using 50Hz. Japan had a closed door policy and when they opened up; they decided to emulate the country best in the respective fields. So Japan followed mechanical engineering from the Germans and electrical engineering from USA. Therefore Japan is one other major country which uses 60 Hz (actually about half of Japan uses 60Hz). Other than Japan, countries close to USA from Brazil up to Canada use 60 Hz.

In a single phase system, load has to be balanced in the housing estate or among machines in a factory such that each of the three phases have almost equal load.

Two forms of connecting three phases are the Δ and the Y connection. In a Δ connected motor, neutral is not connected. In Y connected motor, neutral is usually connected (it can still run if N is not connected).

AC going to a motor in Δ connection is analogous to three pipes sending water to a motor. The water is sloshing back and forth in each of the pipes. The motor uses this 'back and forth sloshing water' to do mechanical work. The phase difference at one moment of time is the voltage each coil has at one moment of time. That phase difference continuously varies from 0V to 415V and back to 0V thereby magnetizing and demagnetizing the coils in the induction motor in a sequential manner. Therefore the different potential between each of the three phases at an instant is similar to a single AC live wire having a higher potential than neutral at one moment of the sine wave and zero potential at another moment in time thereby magnetizing and demagnetizing the single phase motor coils.

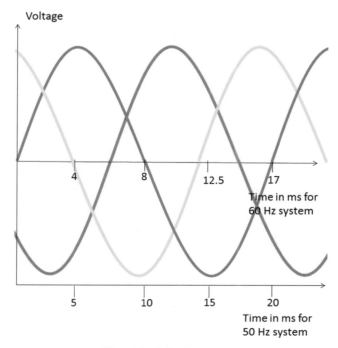

Fig. 15: AC sine waves

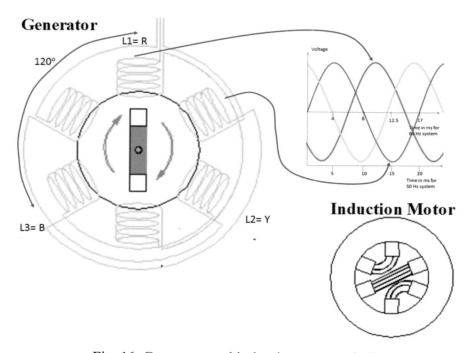

Fig. 16: Generator and induction motor winding

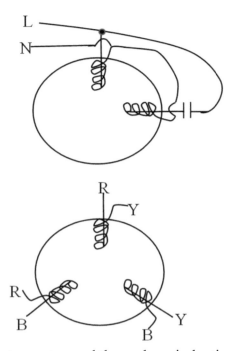

Fig. 17: Single phase motor on top and three phase induction motor on the bottom

As can be seen from Fig. 17 top and Fig. 15, a single phase motor anticlockwise coil connected to L3 (B) phase will have zero magnetic strength at 0ms, a peak magnetic strength at 5ms and zero magnetic strength at 10ms. It will have opposite peak magnetic strength at 15ms and back to zero magnetic strength at 20ms. Actually as shown in Fig. 17, the single phase motor is a modification of the three phase motor. The modification is with the use of a capacitor. The top or vertically placed coil of a single phase motor is energized with L3 (B) phase live wire. L3 (B) phase live wire is connected to the top wire of the motor and this magnetizes the coil as the magnet passes the L3 (B) clockwise coil in the far away power station. Later the magnet in the generator passes the anticlockwise L3 (B) coil magnetizing the fan's bottom coil. That is, initially the magnet is from the top coil to the bottom coil. Then the magnet is from the bottom coil to the top coil. This is a switching magnetic field which should be able to turn the rotor. But the problem is, the top coil and the bottom coil are too far away so a starting coil needed. This is shown as the horizontal coil. The same B phase wire is joined to a capacitor in series before going to the horizontal coil. What will happen is that the capacitor need to be charged up before it releases energy and into the horizontal coil, magnetizing it. Thereby the starting coil becomes a magnet only after the vertical coil becomes a magnet. This gives a starting torque to the single phase motor. The capacitor therefore provided two things a starting torque as well as direction for the motor to turn. This author did try shorting the legs of the capacitor in a single phase motor. The motor does not turn. But if a finger is used to tug the blade in the clockwise direction, the motor turns clockwise. If a finger tugs the motor at standstill in the anticlockwise direction, the motor will continue spinning in the anticlockwise direction. The problem with single phase motors is the capacitor. Capacitors have a relatively short lifespan so single phase motors cannot last as long as equivalently varnished three phase motors. For a couple of years, this author was in charge of the repair center of the Western Digital factory, where all spoilt equipment is sent. In most cases heat is the root cause and the heat normally destroys capacitors. Sometimes a $0.20 capacitor can bring up a $7000 VFD (Variable Frequency Drive). Another thing to note is that all electrical power connection is parallel except for the capacitor in a single phase motor.

Polyphase phase motor was invented by Tesla prior to the usage of polyphase as an AC supply standard. Therefore three phase motor is the original and meant to work well with the three phase supply. Some people do not understand this. For example one Technical manager of a big hotel to which this author was a contractor to wanted to change all the original three phase motors used in the hotel's kitchen blower and exhaust fans to single phase motors. He probably felt this is a simple motor so it should be like his home single phase fan. This author advised him against that because three phase motors is the original stable design.

Looking at the phasor diagram of Fig. 16, if an induction motor has three coils connected to (L2, L3) Y, B phases (L2, L1) Y, R phase and L3, L1 (B, R) phase. L2, L2 (Y, B) coil will have magnetic energy till 2 ms, L2, L1 (Y, R) will have magnetic energy till 5ms and L3, L1 (B,R) will have magnetic energy till 7 ms. Thus as one coil loses magnetism as the phasor voltage waves meet, the other two coils will have magnetic energy to turn the rotor and so forth.

As the magnet in the far away generator passes the L1 (R) phase coil, the coil in the motor in even 1000 km away will be a magnet almost immediately. The magnet then passes the B anticlockwise coil and immediately the B anticlockwise coil in the motor becomes a magnet. Then the generator magnet passes the Y clockwise coil making the clockwise coil in the motor a magnet. Thus the rotor just follows the generator movement.

How fast can the current from the far away generator travel can be calculated approximately as follows, assuming current travels at the speed of light (in bare overhead lines it travels at 97% speed of light):

$$S = \frac{d}{t} \; so \; t = \frac{d}{S} \qquad (17)$$

Assume 1000km distance:

$$t = \frac{d}{S} = \frac{1000km \; X 1000m}{3 X 10^8 m/s} = 0.003 seconds$$

Neutral is connected to ground at the incoming transformer which is why a test pen on neutral does not light up. But it is not safe to touch neutral because the three phases may not be perfectly balanced in places like a factory production floor, resulting in some current in the neutral.

Ground wire (green or Earth wire) on the other hand is connected to copper rods all over the factory, hospital, airport etc., thus it is always at zero potential.

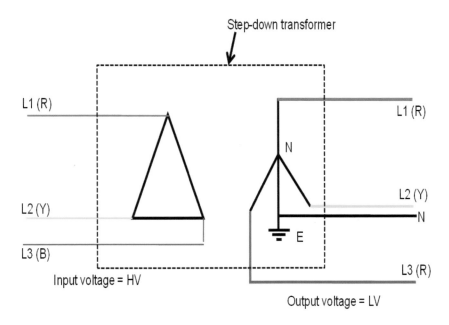

Fig. 18: Transformer substation schematic; typical layout to HV to LV substation

The Fig. 18 shows the schematic of the intake transformer to a factory or home. This intake transformer is normally located in a compound near the electrical installation (like a housing estate) in an area called the Ring Main Unit (RMU) as shown in Fig. 19. In the RMU there is normally a transformer plus a switchgear. An 11 KV line is already running underground. A businessman needs to build a new shop house. So the power company finds the nearest 11kV line and instructs where to locate and build the RMU. The main switchgear is a two individually controlled switches, that is one can ON while the other is OFF or both can ON etc. Coming out of these two fuses is a big ceramic fuse about 4" in diameter and about 14" in length.

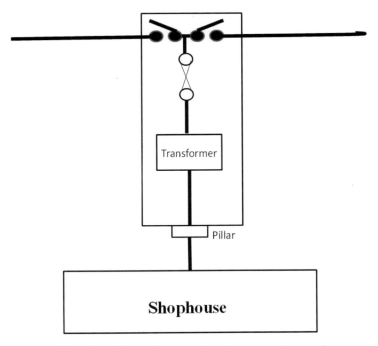

Fig. 19: Substation to shop-house schematic

The three coils L1, L2, L3 (R, Y, B) are looped as a star point and this star point is joined to the Ground (Earth) rod. Sometimes up to 40 Ground (Earth) rods are planted and all these Ground (Earth) rods are joined with a 1" X 1/8" copper tape or 35mm² bare wire. This Ground (Earth) must be good because this is the return point of the all the three phase L1, L2, L3 (R, Y, B) live wires that goes to the load (home, factory etc.). For example if this return path is not as good as possible, the current in the life wire will try to go to Ground (Earth) via human bodies as shown in Fig. 20. But if neutral is very good it will not at all go through a human body even if a human is touching a live wire.

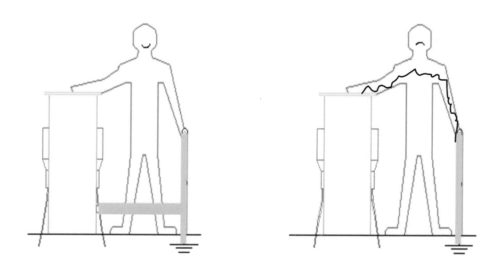

Fig. 20: Man touching live switchgear casing without earth bonding on left
and man touching live switchgear casing with earth bonding

Current takes the easiest path the go to ground will not at all be interested in going through a human body which has resistance. Normally substations transformers the Ground (Earth) resistance is below 1Ω sometimes specifies as 0.5Ω. Generator stations also need such low resistance. The resistance measurement is taken with a Ground (Earth) resistance meter as shown in Fig. 21.

Fig. 21: Ground (Earth) Resistance meter used to measure earth resistance
of a telecommunication tower grounding (earthing) system

This meter has three wires, the green wire is clipped to the Ground (Earth) rod and the other two wires are factory built to a correct length to detect the Ground (Earth) resistance. Thus these two wires need to be pulled as far as possible and at that end it is clipped to a one feet long T metal (Fig. 22) is pushed into the ground.

Eight inches T that comes with Earth Resistance meter pressed into the ground and the yellow wire clipped to it. A similar T is clipped to the red wire. Green wire is clipped to the ground (earth) rod. The yellow and red wire must he stretched to their maximum since their length was calculated by the manufacturer of the meter.

Fig. 22: Earth Resistance meter

The main button on the Ground (Earth) resistance meter is then pushed to get the Ground (Earth) resistance measurement.

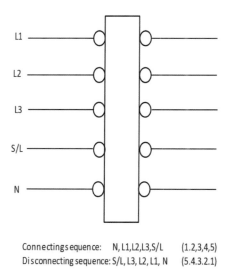

Connecting sequence: N, L1, L2, L3, S/L   (1.2,3,4,5)
Disconnecting sequence: S/L, L3, L2, L1, N   (5.4.3.2.1)

Fig. 23: Jumper connection in overhead lines

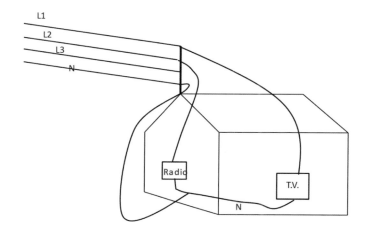

Fig. 24: Problem of not having N connection in three phase home

When connecting live jumpers across section poles as shown in Fig 23, the sequence to be followed is N then L1, L2, L3, S/W (R, Y, B, S/W). S/W stands for Switch Wire which is used to power the street lights. The reason for this is say, L1 (R) and L2 (Y) is connected as shown in Fig. 24, without N being connected. Now the TV in this three-phase home is connected to L1 (R) phase and the radio is connected to the L2 (Y) phase. Now L1 (R) and L2 (Y) are joined via the N wire giving $120\sqrt{3} = 208V$ ($240\sqrt{3} = 415V$ British) powering the TV and radio. The TV and radio is supposed to receive only 120V (240V British) so they will be damaged. Basically the White (Black British) N wire has become a live wire. So with the procedure of connecting N first is designed to provide 0V in N wire first before L1 (R) and L2 (B) are connected thereby the TV and radio will not blow up. N is basically Ground (Earth) at the substation. In some Australian farms which are up to 100kM length and breath, it would be expensive to run L and N wire so only L is run to the house and after the load (say a TV) the return path is E rod of the home. Basically as mentioned before electricity only wants to go to the ground and if some equipment is placed in-between source and Ground (Earth), that equipment will function.

Chapter 7
Line voltage and phase voltage

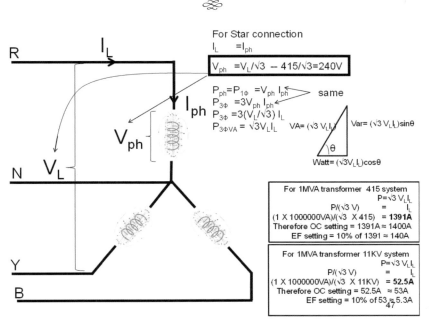

Fig. 25: Star connection

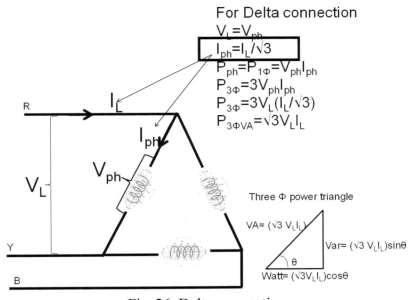

Fig. 26: Delta connection

A three phase system has three phase-to-neutral voltages that are most of the time equal in magnitude and displaced by 120 electrical degrees; because the generator coils in-between one phase's clockwise coil and another phase's clockwise coil are physically 120° apart. The voltages in each phase are often referred as **phase voltages** but phase-to-neutral voltage is a more accurate term. The phase voltage is written as V_p in Fig. 25 and Fig. 26 which are the two methods of joining the phases up.

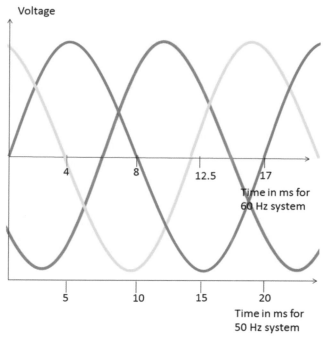

Fig. 27: Three phase waveform

A three-phase system also has three phase-to-phase voltages that are also most of the time equal in magnitude and displaced by 120 electrical degrees. These voltages can either lead or lag the phase-to-neutral voltages depending on which phase-to-neutral as can be observed in Fig. 27 above. A voltage must be chosen as reference to consider it leading or lagging. Phase L1-L2 (R-Y) voltage is numerically equal to phase L2-L3 (Y-B) voltage, but they are 180 degrees out of phase with each other. The phase to phase voltage is generally called line voltage.

Chapter 8

Conductors and Protection

The function of electrical protection is mainly to safeguard the conductors because if conductors burn homes or buildings will burn. Copper is the most common electrical conductor. It has a melting point of 1083 centigrade, an atomic radius of 1.57 Angstrom (10^{-10} m). The outer shell of any atom is the valence shell, for Cu the valence shell has one electron as shown in Fig. 28.

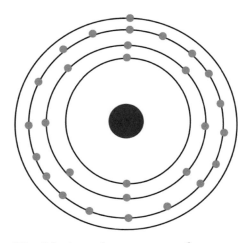

Fig. 28: Atomic structure of copper

With the help of STM (Scanning Tunneling Microscope), humans can actually see the atom today. The inventor of the STM received a Nobel Prize because it took all humanity one step ahead in being able to see the atoms which were theorized long ago.

An ion is an atom or molecule in which the total number of electrons is not equal to the total number of protons, giving it a net positive or negative electrical charge. An **anion**, from the Greek word ano, meaning 'up', is an ion with more electrons than protons (**negative charge**). A **cation**, from the Greek word kata, meaning 'down', is an ion with more protons than electrons (**positive charge**).

It should be noted that all ions even ions of plastics are conductors of electricity. Ionic gas is defined as plasma. So ions should be kept away from electrical systems as far as possible. The specification states that if the MSB (Main Switch Board) room is underground there should be a fan at this window (forced ventilation). MSB rooms create lots of ions especially at the capacitor bank. Here high powered capacitors are continuously being switched on and off (according to the amount of inductive load present) resulting in arcs in the high power contactors. These arcs convert atoms of the air in the MSB room into ions. So if there is no ventilation eventually the MSB room will be filled with ions. Then one day when there is a fault in the building and the ACB (Air Circuit Breaker) need to operate, conductive ions instead of insulator atoms are blown in-between the ACB contacts resulting in an explosion.

In the power company that this author worked for, one technician got second degree burns because he switched the Oil Circuit Breaker (OCB) whose indicator showed less than normal oil level. When insulator oil is blown across the contact points the arc is blown to become a long U shaped arc. So when less oil is blown the length of the U shaped is insufficient and this caused an explosion. So it is very critical that the breaker works exactly as specified.

This idea of blowing an arc is used from the smallest MCB (Miniature Circuit Breaker) in the home to the largest breaker of big power companies. Fig. 29 depicts a typical home MCB.

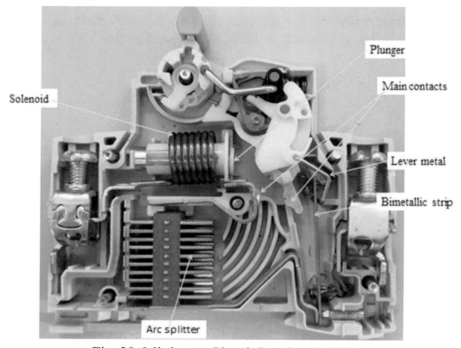

Fig. 29: Miniature Circuit Breaker (MCB)

The mechanism above causes a blow of air in-between the silver plated main contacts. This will cause the arc to go towards the Arc splitter which splits or breakup the arc into it's groves. In a MCB there are two systems, one is overload protection and another is short circuit protection. For the 10A Schneider MCB in Fig. 29, the overload protection value is written on the front as 10A and the short-circuit protection is written as 4500A. This means this circuit breaker will trip if a motor bearing jams drawing more than 10A but it can still trip (and not blow up) if up to 4500A short-circuit current flows through it. Note if the MCB blows up and separates the contacts, it is very good but if it blows and fuses the contact to join, it is very dangerous for the equipment in this line and thereby the whole installation can potentially go up in flames.

Looking at Fig. 29, the overload protection is performed by the bimetallic strip. The bimetallic strip is made of copper and iron pasted together. Copper has a lower melting point compared to iron so it expands faster thereby bending the bimetallic strip to the right. As it does so, it pulls the Lever metal which pulls the white plastic, this slight movement will be further continues by a spring to open the normally closed Main contacts. As this happens the mechanical design of the MCB blows air in-between the Main contacts causing the arc to be sprayed to the arc splitter thereby quenching the arc. Note the two mechanisms of arc quenching; by bowing the arc it becomes thinner which is safer than a thick arc, then this thin arc is split by the arc splitter. The other end of this thin arc need not go to the other silver Main contact; it can finish up at the copper below the Arc splitter, thereby not damaging the silver of the Main contact. Note the copper below the arc splitter goes all the way to the left because the arc may in a fraction of a second reach somewhere there. The arc emanating from the first silver coated Main contact on the left is not as damaging as the arc being shot at the second Main silver plated contact on the right. So it is designed to be shot at the copper below the Arc splitter.

The same mechanism with little modification is used in giant breakers. Of course giant breakers do not depend on fresh air to be blown in-between the Main contacts; insulating oil or increasingly SF_6 gas is used. One of the modifications in giant breakers is to add an arching contact made of tungsten. Note tungsten is the material used to make filaments of Edison bulbs so they are hardy. This arching contacts project out longer and sharper because electricity always like longer and sharper points so all the arc will chose to go to this arc-withstanding tungsten and not the silver plated Main contacts. By the time the silver plated Main contacts joins there will be little or no arc.

The short circuit protection happens when thousands of amps flow through the solenoid, magnetizing it and extending the Plunger to the right. This small push will cause the spring to complete the opening of the Main contacts. Note that a short circuit of even thousands of amps will not bend the bimetallic strip. It takes about 30 seconds to a minute for the bimetallic strip to bend which is good enough for 12A flowing to a motor instead of the rated 10A (an overload).

When purchasing a MCB there are three categories, Class A, B and C representing 5X, 10X and 20X short circuit triggering respectively. Class A enables tripping at 5X the overload rating on the MCB. For example a 6A rated MCB will trip with a short circuit of 5 X 6 = 30A flows through. Class B will trip with a short circuit of 10X (10 X 6 =60A) and a Class C will trip at 20X (20 X 6=120A) rated current. Class A, B and C are only related to short-circuit but the overload mechanism for all three classes are the same. Of course the bimetal for a 40A MCB is thicker than a 6A MCB. The coils of the solenoid only have a few turns for a 40A MCB and lots of turns in a 6A MCB. 40A – 60A is a huge current to appear in this simple MCB, note 1A is enough to kill a human.

The reason for Class A, B and C are for example Class A is meant for light bulbs. An Edison bulb filament is cool initially so the atoms (or ions) are not moving much so the resistance is low. Note if the ions are not moving at all there will be zero resistance as in a superconductor. So current can easily flow through the cable resulting is a very short period of time of almost short circuit. We do not want the MCB to trip because of this. Note the tiny span of time of very high ampere will not bend the bimetal of the MCB; it needs about 30 seconds to a minute of current flow to significantly bend the bimetallic strip enough to move the lever metal. Class B are for motors where inrush current upon starting can be many times steady state current. Class C is meant for big motors where inrush current is proportionately larger.

From a good quality MCB all the way to giant switchgears, the contact points are silver plated (silver being the best conductor). This author uses Schneider because the three bigger electrical part supplier, GE, Siemens and ABB do not have an office in his hometown. Schneider has an office and at least can inform customers which electrical shop is selling original Schneider products. The office itself does not sell the products. There is imitation Schneider in this author's home town being sold by other electrical shops. But as an electrical contractor this author never sacrifices in brand. A world top company has a reputation to keep. And a contractor cannot blame the parts. Therefore it is the contractor's job to ensure the parts are of the very top quality; and there are some very terrible brands out there in the market. So the job of a good electrical contractor is to continuously keep up with who are the best international suppliers and also who are the best local suppliers. Big international suppliers do fail over time. Case in point is Enron or Arthur Anderson both of which do not exist anymore despite being world giants prior to scandals. For local suppliers, it is the experience of this author that if the owner of the shop is still in the shop it should be more reliable. Owners, who dare to face disgruntled customers by being in their shop and not delegate everything to sales personnel, should be better at selling genuine electrical products.

The latest breakers blow sulfur hexafluoride (SF_6) in-between the contacts. In most common Oil Circuit Breakers (OCB) the oil can eventually get ionized with many switching. Note sending an arc through an element or molecule is the method of converting atoms to ions. Cooling the ions is the method of changing it back to atoms. Hence the high pressure oil in-between the contacts to cool the region and convert some of the ions formed by the arc back to atoms. The ionizer air-conditioner used in some homes has a small arc through which air is blown to create ionized air. Therefore eventually the oil in OCB will be ionized and instead of blowing insulator oil in-between the contacts, a conductor (ionized oil) is blown and an explosion will occur. It is very critical that the blowing of the arc happens with sufficient pressure and there is an arc splitter to quench this arc; every portion of this process must occur or else there will be an explosion. Recently there was a case of a technician observing a low oil level in an 11kV switchgear but decided that it was sufficient and switched it to get an explosion and second degree burns all over his body. It is just like a car engine, the car cannot run without the engine oil. This author had an experience of hitting a piece of metal on the road which punctured a hole in the engine oil filter but continued driving a few miles, then the engine stopped dead. When the mechanic opened up the engine, there was massive damage all over the engine and all parts except the engine block had to be replaced; the engine block had to be re-bored. When this author worked for a power company all the blown up OCBs were displayed as a warning to technicians to take precautions when operating them. The metal body of the breakers is 5-7 mm thick (like the iron of military tanks) but it blows up and is twisted like thin paper; such is the power of electrical explosions.

The correct procedure for technicians is to drain out a little oil at intervals specified by the manufacturer. Typically it is once in five years. Most staff of power companies use time based PM (preventive Maintenance) but it should be after a certain number of switching. Probably the hassle with record keeping is a deterrent to this procedure. The more times arching occurred through the switchgear oil, the more often the oil should be tested for ionization level; so a better frequency should be the number of times switching occurred. The equipment to test ionization level of the oil is called a Dielectric Tester. Higher ionization levels of the switchgear oil the less the dielectric strength of insulation capability. The word dielectric which originated as the material in-between plates of capacitors is now generally used to replace the word insulator in electrical power engineering books and journals.

The act of blowing air, SF_6 or oil in-between electrical contacts does three things:

1. The arc gets thinner.
2. The arc is split by many window-like insulators called arc-splitters at the outlet.
3. The material (oil or gas) blown in-between the contacts gets deionized by the cooling action of the blowing. But it should be noted that blowing compounds in-between arcs is the way to create ions in the first place. Ionizer air-conditioners or general home ionizers have air being blown across arcs to produce ions. This author does not recommend either of these two products.

In the compound SF_6 the sulfur and the six fluorine atoms have a very strong bond. Thus SF_6 is unreactive and even very high arcs cannot split the F from the S. Sulfur and fluorine are easily available cheap elements so countries that make SF_6 need not be advanced countries, anyone can take easily available sulfur and fluorine and heat it up in a container to make SF_6. It is harmless to the human body because it is so unreactive but as a global warming gas it is 22,800 times more dangerous than CO_2; having a lifetime in the atmosphere of 800–3200 years. Therefore the United Nations is watching out for the usage of this gas. The power company this author worked for now uses lots of SF_6 breakers from ABB and Siemens and the technicians report that the SF_6 do leak out. This made this author conclude that human technology is not capable of placing a gas in a container for long periods of time. This conclusion can be made because Siemens is the world's number two electric equipment supply company and ABB is the world's number three electric equipment supplier company (GE is number one). If these over 100 years old company cannot keep gas in a container who can?

The very latest RMUs developed by Subhashish Bhattacharya and team of NC State University, USA as a collaborative project by the university, CREE and ABB. This RMU is very small due to it's utilization of the latest development in IGBT (Insulated Gate Bipolar Transistors). The IGBT can switch at 700 ns for a 6.5kV and 1070 ns for a 10kV one.

Note the following data:

IGBT switching timing is 1070ns = **1.07×10^{-6}s**.

Oil Circuit Breaker (OCB) switching timing = **1×10^{-1}s**.

Thus the IGBT **switches 100,000 times faster** than the OCB. It is because of this that IGBT used as switchgear can by-pass the requirement to blow a liquid or gas in-between the contacts. This causes the redundancy of lots of arc quenching parts thereby making the switchgear very small.

A 1 MVA transformer in this RMU is only 25" X 10" X 15" (250mm X 240mm X 390mm). This small size is enabled because this transformer is operated like a SMPS (Switch Mode Power Supply) where the higher the incoming frequency the smaller the transformer. The main advantage of this system is switchgears will not explode and kill people anymore. This author has a close Malaysian friend while studying at South Dakota State University, USA. He graduated and became an engineer in the main electric company in Malaysia. A year later he and three others were killed because the switchgear he was operating blew up. This IGBT RMU can prevent such fatal accidents. Other obvious advantage for places on earth where land is very expensive is that land need not be purchased to locate the RMU. The RMU is so small that it can be located in a small section of a home (power company can rent this space from a house owner etc.).

Some MSB rooms are designed with air-conditioning which just recycles the existing air. Electrical and electronics in the MSB last longer with air conditioning but even a small 1' X 1' window is essential to allow the ions to escape to the atmosphere and new atoms to come in. In a MSB ions are created mainly at the Capacitor Bank unit where continuous switching of MCCB is done automatically by the Power Factor Regulator to correct the power factor (PF) of the installation. The window of the MSB room should be covered with a netting preferably metallic netting so that animals especially lizards cannot go in. This author has seen holes made in fiber glass nettings made by lizards. This author did a wiring job in a goat farm in a rural area where every single live point has a dead lizard. Lizards seem to like the relative warmth of live points. The Distribution Board (DB) needs to be totally sealed up in a rural setting. Lizards seem to the most hazardous animal for electrical problems.

The power company statistics indicate the biggest culprit in tripping of overhead lines are squirrels because they are straight and long between the nose and the tail. Sometime snakes do climb up power lines and monkeys get electrocuted as they hold two phases overhead wires.

In the MSB, before each capacitor is a MCCB (Molded Case Circuit Breakers) and then a high power contactor. MCCB handles currents from >100A to 1000A. A MCB (Miniature Circuit Breakers) of homes handle current from 1-100A. Arcing in contactors and all breakers results in the atoms being converted into ions. Cooling can convert the ions back to atoms. Other names for ions are plasma and free radicle (used by doctors). Doctors will tell you that lots of free radicles in your body can inflict damages to your health. This is basically ions. This author does not know how ionizer air-conditioner or ionizer refrigerators got approved. Being a Radiation Protection Officer (RPO) for over 20 years it is known that it is not nuclear bomb rays (α, β, γ, n) that kill but the second effect of the ray ionizing body atoms that kills people. Ionizer air-conditioner will clean rooms of germs by killing them but the same ions can kill humans cells. Note we have more germ cells than human cells in our body anyway, if specifically killing germs is the advertised purposed of these air-conditioners.

The other protection device used in homes is the RCD (Residual Current Device). Previously this was called GFCI (Ground Fault Circuit Interrupter) or ELCB (Earth Leakage Circuit Breaker). While the GFCI (ELCB) detects fault current in the Ground (Earth) wire, the RCD detects a difference in the current value in L and N wires as shown in Fig. 30. The L wire will have a magnetic field around it following the Right Hand Thumb Rule. The N has returning current and therefore has a magnetic field turning in the opposite direction compared to the L wire. These two magnetic fields oppose each other resulting in no current generation in the detecting CT (Current Transformer) coil. But if a person touches a L wire, he will leak some current to the Ground (Earth) via his feet resulting in less current going back via the N wire. Therefore the magnetic field of L wire and N wire will not cut out each other and a little (the word Residual of RCD indicates this little) AC magnetic field will energize the CT. Current from this CT and will trigger an amplifier circuit to energize the solenoid which will break the circuit. Single phase RCD are set to trip at 100mA difference between L and N.

This situation can be compared to a water pipe sending water to a home and another going out of the home (note in actual water pipes in homes it is just one way into the home, with a dead end). It this water pipe water is used to turn a blade which is connected to a fan, we have a hydraulic energy fan in our home. Say the home has many such devices driven on this hydraulic energy; this is equivalent to the typical electrical devices in homes. The incoming and outgoing pipe to the home has the same pressure; an RCD that detects the difference in incoming pipe and outgoing pipe sees no difference. But if there is a faucet in the home, there is a leak to ground (earth) and now the outgoing pipe will have a different pressure compared to the incoming pipe; this is equivalent to a human touching a live wire.

Fig. 31 is a three phase RCD. In this case the current additions of the L1, L2, L3 (R, Y, B) phases is -5A +3A -4A = -6A. So the outgoing N wire will carry +6A. A CT placed surrounding all four wires detects nothing. But if a human touches any of the L wires, he will leak some current to ground (earth) through his feet and now the three phase combined input to the house is not the same as the N output from the home. A CT around all four wires will detect this current difference and trigger a break of the circuit. Three phase RCD are set to trip at 300mA difference between the three phase combined (which goes into the home) and N (which comes out).

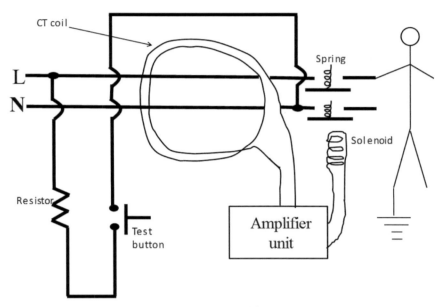

1Φ RCD is set to trip at 100mA difference between L and N

Fig. 30: Single Φ RCD schematic

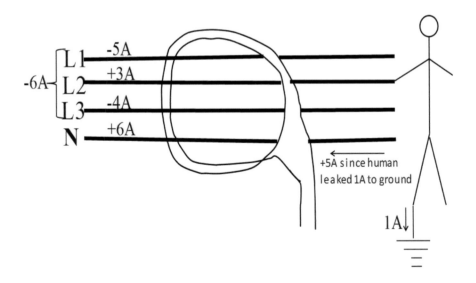

3Φ RCD is set to trip at 300mA difference between L1, L2, L3 combined and N

Fig. 31: The Φ RCD schematic

Table 1: Resistivity of various conductors

Resistivity of some metals

Material	Resistivity (ρ=Ω·m) at 20 °C	Melting point	Oxide melting point
Silver	1.59×10^{-8}	962°C	
Copper	1.72×10^{-8}	1083°C	1200°C
Gold	2.44×10^{-8}		
Aluminium	2.82×10^{-8}	660°C	2000°C
Calcium	3.36×10^{-8}		
Tungsten	5.60×10^{-8}		
Zinc	5.90×10^{-8}		
Nickel	6.99×10^{-8}		
Iron	1.00×10^{-7}		
Tin	1.09×10^{-7}		
Platinum	1.06×10^{-7}		

Soldering lead is 60% tin and 40% lead,   Si boiling point is 2357°C

Table 2: The Periodic Table

1 H 1.008																	2 He 4.003
3 Li 6.941	4 Be 9.012											5 B 10.81	6 C 12.01	7 N 14.01	8 O 16	9 F 19	10 Ne 20.18
11 NA 22.99	12 Mg 24.31											13 Al 26.98	14 Si 28.09	15 P 30.97	16 S 32.07	17 Cl 25.45	18 Ar 39.95
19 K 39.1	20 Ca 40.08	21 Sc 44.96	22 Ti 47.88	23 V 50.94	24 Cr 52	25 Mn 54.94	26 Fe 55.85	27 Co 58.93	28 Ni 58.69	29 Cu 63.55	30 Zn 65.39	31 Ga 69.72	32 Ge 72.61	33 As 74.92	34 Se 78.96	35 Br 79.9	36 Kr 83.8
37 Rb 85.47	38 Sr 87.62	39 Y 88.91	40 Zr 91.22	41 Nb 92.91	42 Mo 95.94	43 Tc 98.91	44 Ru 101.9	45 Rh 106.4	46 Pd 107.9	47 Ag 112.4	48 Cd 114.8	49 In 118.7	50 Sn 121.8	51 Sb 121.8	52 Te 127.6	53 I 126.9	54 Xe 131.3
55 Cs 132.9	56 Ba 137.3	71 Lu 175	72 Hf 178.5	73 Ta 180.9	74 W 183.8	75 Re 186.2	76 Os 190.2	77 Ir 192.2	78 Pt 195.1	79 Au 197	80 Hg 200.6	81 Tl 204.4	82 Pb 207.2	83 Bi 209	84 Po 209	85 At 210	86 Rn 222
87 Fr 223	88 Ra 226	103 Lr 262.1	104 Rf 261.1	105 Db 263.1	106 Sg 263.1	107 Bh 264.1	108 Hs 265.1	109 Mt 268	110 Uun 269	111 Uuu 272	112 Uub 277	113 Uut	114 Uuq 289	115 Uup	116 Uuh 289	117 Uus	118 Uuo 293

57 La 138.9	58 Ce 140.1	59 Pr 140.9	60 Nd 144.2	61 Pm 146.9	62 Sm 150.4	63 Eu 150.4	64 Gd 152	65 Tb 157.3	66 Dy 158.9	67 Ho 162.5	68 Er 167.3	68 Tm 168.9	70 Yb 173
89 Ac 227	90 Th 232	91 Pa 231	92 U 237	93 Np 238	94 Pu 244.1	95 Am 243.1	96 Cm 247.1	97 Bk 247.1	98 Cf 251.1	99 Es 252.1	100 Fm 257.1	101 Md 258.1	102 No 259.1

As can be seen from Table 1, silver is the best conductor of electricity. Copper is number two and aluminum is number three. Table 2 is the whole Periodic Table and depicts the location of the commonly used electrical elements like copper (Cu), aluminum (Al), silver (Ag), iron (Fe) in transformer and motor core, gold (Au) used in electronics, silicon and surrounding elements for electronics. There is a great shift in the electrical industry from copper to aluminum mainly due to cost. The price of copper is raising everyday but aluminum price is decreasing every day because it is the third most abundant element on earth after oxygen and silicon. It is found as bauxite and with lots of electricity, electrolysis of molten bauxite produces pure aluminum. But actually aluminum is not at all optimum for electricity. It forms a layer of oxide (white in color) in microseconds and all oxides of metals are insulators.

As can be seen from the Table 1, aluminum's melting point is about half of that of copper. So an energized join of two aluminum cables will be arcing because electricity cannot flow easily through the aluminum oxide in-between. Arcing will lead to fires and aluminum burns at half the temperature of copper. So aluminum is already a bad conductor because it forms the oxide layer in microseconds which will lead to arching and it also burns at half the temperature. When power companies do overhead installation with aluminum conductors their staff are required to use torque wrench to bolt the joints very tight such that metal of the bolt and conductor fuse and oxides cannot form easily in-between. For copper conductors, this author has done wiring in his parent's 47 year old home where it can be seen that the stripped copper is still perfectly of copper color; which means the copper wire has not corroded. Corroded copper is black and later turns green and blue as it reacts with sulfur and carbonates. If such corroded wire is observed, it must be filed off. Filing seems the best solution as sandpaper is hard to control and therefore will tend to sand our fingers in the process. Alternatively a penknife can be used to scrape off the black copper oxide (CuO) layer. In the picture in Fig. 32, left is the original blackened wire. This was cut and stripped but the wire inside was still blackened with oxide. So a penknife was used to scrap the black oxide away.

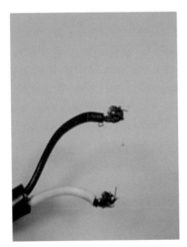 

Fig. 32: Removing corrosion with penknife

So in the electrical industry all technicians, engineers and scientists are continuously fighting the problems of oxides of copper and aluminum (corrosion). But metal oxides are used in the electrical industry in two cases as follows.

In surge arrestors, zinc oxide (ZnO) is placed in-between the Live wires and Ground (Earth) wire which goes to the Ground (Earth) rod as shown in Fig. 33. In a Distribution Box (DB) the surge arrestor is placed at the power incoming portion so as in Fig. 34, it should be placed before the RCD. If a lightning strikes any of the Live or Neutral wires, the bond between zinc and oxygen will break and the zinc will conduct the surge into Ground (Earth). So it takes a surge as from lightning to make the conversion from ZnO to Zn. It must be noted that the world top electrical companies are GE, Siemens, ABB and Schneider. This author met two contractors that purchased surge arrestors costing thousands of dollars while Schneider sells it for under a hundred dollars. A common sense is that how can a company ranked below Schneider can make a product and sell it for far higher cost than Schneider; it is simple logic that these two contractors got cheated.

The other use of metal oxide is in mineral insulated cables or MI cables for short. These wires power boilers so plastic (PVC or XLPE) wires cannot be used because they will melt. In MI cables, magnesium oxide (MgO) is used to separate the three live wires and N; no plastic is used. Other than it's electrical insulation properties, MgO also has a heat resistance property, thereby providing heat insulation to the copper conductor from the boiling load for example. Looking at the two compounds, ZnO has a weak bond between Zn and O which could be broken by a surge of electricity.

It is sometimes difficult to remember which oxide is used in surge arrestor and MI cables. A way to remember is when we were young, our school science teacher put a small piece of magnesium over a Bunsen burner and it burnt with a blinding bright white flame. Note the fire triangle; one side is fuel, the other heat and the third is oxygen; all three is required for fire. The intense burning with bright blinding light indicates that the fuel (magnesium) is very reactive. This burning is oxidation or adding oxygen to magnesium for in other words the formation of corrosion. If magnesium burns so intensely it's bond with oxygen is very strong and even high voltage cannot split the magnesium ion from the oxygen ion. So it is used as the insulator in MI cables and to give extra insulation in most high voltage underground cables.

Zinc on the other hand will not burn or oxidize with a school science lab Bunsen burner. A very high temperature will be needed for it to burn. In fact it is one of the least reactive metals so it is the element used to coat iron to prevent rust. The resulting iron is called galvanized iron. So the weak bond in-between zinc and oxygen is utilized in surge arrestors as shown in Fig. 33. The surge arrestor is the incoming protective device of newer home and offices. Most home has the incoming device as the RCD as shown in Fig. 34. In some countries the incoming device is the isolator. The other protective device is the MCB (Miniature Circuit Breaker). So of the three protective devices only the isolator is a dumb device; it is a manually operated device without intelligence. RCD trips if there is a Ground (Earth) fault and MCB trips due to too much current being drawn. The new surge arrestor should be placed as the incoming device. In a single phase home, L and N go into the top of the device and then is looped to the RCD. At the bottom of the surge arrestor is a green wire going to the Ground (Earth) bar and thenceforth to the Ground (Earth) rod. In normal case, the zinc oxide acts as an insulator in-between the incoming and the Ground (Earth) rod. But when lightning strikes, the Zn easily dissociates from the O and lightning surge travels through the Zn to the Ground (Earth) rod; thereby protecting the home.

3Φ   1Φ

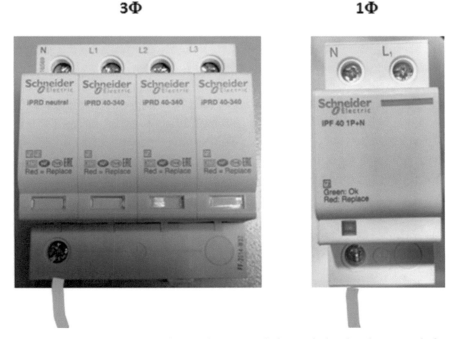

Fig. 33: Surge arrestor, three phase on right and single phase on left

Fig. 34: Three phase Distribution Board

Soldering is the best way to join wires. This author do not like crimping unless in some projects where it is a requirement. Crimping stresses the crimp lugs. In mechanical engineering, it is a standard practice to burn the metal after stressing it. For example in making a car bonnet, the sheet of metal is first clamped than heated to relieve the stress. Crimp lugs are not heated after stressing it with a crimping tool so stress remains in the lug and wire causing fast corrosion and deterioration of the join. On the other hand in soldering, two wires are twisted (stressed a little) then soldered which is a heat treatment to relieve stress plus enable a very strong joint. If the two sides of the join are pulled hard enough, the wire will not break at the join but at the copper portion, such is the strength of the join. It must be noted that in making joins of copper wires, lots of twists must be made to ensure a proper copper-copper connection. The solder is applied outside, mainly for mechanical strength. If the copper-copper connection is good enough due to the many rounds of twisting, the current will avoid flowing through the solder (electricity always takes the easiest path). Of course the join should not be twisted too much, it is a skill to be able to realize how many twist a wire can withstand without breaking. Solder should never be allowed to get in-between the copper. That is, never have copper-solder-copper join. This is because solder is 60% tin and 40% lead. As can be seen in Table 1, tin is number 10 down the list in conductivity while lead is way down. Therefore solder is a bad conductor of electricity so a copper-solder-copper join will be a heat point and eventually an electrical fire point. Solder is a good mechanical fastener. If a soldered join is pulled, the soldered point will not break as fast as the solid copper potion. So the lots of twisting are not for mechanical strength but to ensure as big a surface area as possible of copper-copper connection. Soldering can be done with a gas soldering iron. If an electric soldering iron is used to solder a live wire, it will trip the RCD. If a gas soldering iron is not available or if your gas container is empty, the last resort is to use a candle, it is fast and effective except that there is lots of carbon formed in the process, this carbon can later be wiped out. So always keep a candle in a good container in your toolbox. Without a container, it will mess up your toolbox. Be careful when soldering a live wire many people forget the solder lead they are placing on the live wire is a conductor.

In the experience of this author who has a technical school, there is a three phase induction motor in the school as shown in Fig. 35 below:

Fig. 35: Autotrans motor circuit with motor terminal connection without crimp lugs

The termination at the six motor terminals used to be crimped. But in a technical school the motor incoming wires are constantly moved many times a day; the crimped join to the motor will last a maximum of two days before it breaks. One day one of the students showed a way to connect without crimping as shown in Fig. 35. A wire stripper was used to move up the PVC about 1.5 cm, the strands of copper separated about equally (left and right having about equal number of strands) and through this hole, the motor termination screw is moved in and the nut is turned on the screw. There is very little stress to the wire, no soldering is needed. Once this author adopted this method, plus tying the incoming wire with a cable tie via a screw hole, the motor termination of all six wires did not break for three years (crimping join lasted for maximum of two days and this method lasted three years).

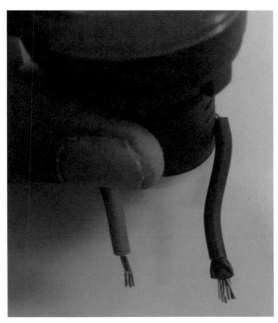
Fig. 36: Screwing on PVC insulation

Another common problem in wiring is screwing onto plastic insulation as shown in Fig. 36. This author believes this is caused by many technical schools instructing students to ensure the copper wire cannot be seen. So the students strip as little as possible and therefore have a higher chance of screwing onto the PVC (polyvinyl chloride) or XLPE (cross linked polyvinyl chloride) insulation. One 20 year veteran of the electric industry said is the most common problem he has to repair. Equipment will work initially as there is little spot of copper conduction at the edge of the PVC. But this spot is of course not sufficient to carry the designed current so it will heat up as the atoms within it vibrate not allowing the electrons to flow easily. This heating up will cause the atoms to vibrate even more causing even more resistance to electron flow. Eventually if a serious fire does not break out, small arcs will form oxides which will eventually render the conductivity through this join to be zero.

The resistance R of a conductor of uniform cross section can be computed as:

$$R = \frac{\rho L}{A} \qquad (18)$$

where

L is the length of the conductor, measured in meters

A is the cross-sectional area, measured in meters square

ρ (Greek: rho) is the electrical resistivity (also called specific electrical resistance) of the material, measured in Ohm · meter. Resistivity is a measure of the material's ability to oppose electric current.

Some people this author met asked won't the fuse prevent damage from a short circuit. Actually most fuse are designed to protect for overload conditions; meaning if a motor is supposed to draw a maximum of 4A, a 5A fuse is used. If there is a bearing jam, the motor will draw more than 4A to overcome the jam and if it draws more than 5A, the fuse will break. A short circuit is different. If L and N touches at the wire going to the motor, it is a short circuit. Using the formula below:

$$V = IR \qquad (19)$$

$$\frac{V}{R} = I$$

$$\frac{240V}{0.03\Omega} = I$$

$$8000A = I$$

So the short circuit of L to N resulted in a current or 8000A while the normal operating current of this motor is 4A. If a protective fuse of MCB do not trip in time, 8000A will do some serious damage; most probably a fire. This is the reason all MCB has two ratings. For example the lowest home MCB may be rated 6A which is the overload rating. There is another rating on the MCB like the number 6000 in a box, this is the short circuit rating. The 6000 means this MCB can trip even if a short circuit of 6000 A flows through the MCB. But standard household MCB is a mechanical device and mechanical speeds seems fast for a human being is extremely very slow compared to almost the speed of light speed of electricity. If the 8000A flows to the motor due to a short circuit, even if the MCB does trip, lots of damage could be caused especially if the equipment being run is an electronic one, like an inverter controlled motor. The standard house-hold fuse is also slow to blow to prevent surge from going into the equipment. This author has seen $100,000 equipment being damaged by a short circuit in a ballast of a florescent tube. A standard practice in industry is to protect expensive equipment with a HRC (High Rupturing Capacity) fuse. High in HRC stands for high current as from a short circuit or lightning surges. The top of the line HRC fuse is made by Bussmann. Most big industrial companies trust Bussman. A Bussman fuse the size of the one in a standard plug-top which cost $0.05 can cost $60. But in this author's experience it is better to source it directly from Bussmann because the profit margin a cheater can make by purchasing a $0.05 fuse, coloring it with Bussmann symbols is so high, someone may already have a business doing it.

Another part electricians must watch out for is carbon brush. An original carbon brush for a Hitachi or De Walt drill may cost $20 while the cheapest carbon brush can cost $0.25. There is no way an electrician can determine if the $20 carbon brush he bought is actually the real thing unless they source it from the original supplier (Hitachi or De Walt). This author got cheated once while working in the Western Digital factory. A super long lasting carbon brush from Bodine Motors was faulty but a local supplier supplied the $0.25 carbon brush with all the marking of Bodine Motor carbon brush.

Chapter 9
Motors

Induction motors

The induction motor is an ingenious invention of Nikola Tesla. In a typical large generator, a negative high current initially generated as the DC magnetized rotor passes the L1 (R) coil then it passes the L2 (Y) coil and then in the L3 (B) coil respectively. In the faraway (maybe 1000 miles away) induction motor, located in a population center, a magnetic N is generated first in the L1(R), then in the L2 (Y) and then in the L3 (B) coils. The induction motor stator coils thus get magnetized almost as soon as the respective coil in the far away generator coil gets magnetized. As shown previously in equation (11) below it is about 0.003 seconds later if bare overhead lines are used. In effect the magnetization of the induction motor stator coils follows exactly the rotation of the generator rotor. This creates a whirlwind of magnetism as Tesla termed it.

$$S = \frac{D}{t} \qquad (20)$$

Where S= speed, D=distance, t=time, so

$$t = \frac{D}{S} \, or \qquad (21)$$

$$t = \frac{1000KM \; X1000m}{3X10^8} = 3.3X10^{-3} = 0.00333 seconds$$

Note putting insulators over the conductors reduces the speed of flow as heat builds up around the enclosed conductor and it begins to shape like a longitudinal capacitor (gains capacitive inductance).

The rotor of the induction motor will keep up turning with this whirlwind but with a speed lag named slip. In synchronous (same speed) motors there is no slip meaning the rotor turns at the same speed as the whirlwind. Most work done in industry is done with induction motors mainly because it requires no carbon brush to send electricity to the rotor. Synchronous motor needs a carbon brush to send DC to the rotor. Therefore induction motors can be made to last up to sixty years, as long as the varnish around the copper coil and the two bearings that hold both ends of the rotor are of good quality. The most modern high speed induction motor uses Active Magnetic Bearing (or AMB). Here the shaft is levitated in space by actively controlled electromagnets leaving zero contact in-between stator and rotor.

In the stator of an induction motor, the three phase coils are arranged as in the generator of a power station. Thus a rotating magnetic field (RMF) is created in the stator of the induction motor that rotates synchronously with the rotor of the far away generator.

The interaction of one particular bar of a squirrel cage is studied with respect to the rotating magnetic field in the stator. In slow motion, at the start, the RMF is switched on as someone switches on the three phase induction motor. The RMF field lines will cut by the stationary bar of the squirrel cage rotor. By the right hand Fleming's rule, a current will be induced in the bar. This current will produce a magnetic field using the right hand thumb rule. This magnetic field will lock with the rotating magnetic field (RMF) and the rotor will start turning, to eventually approach the same speed as the RMF (synchronous speed). In this process it cuts more lines of magnetic flux lines from the stator thereby gaining more current flow and therefore increasing magnetic flux lines around itself. The bar will eventually catch up with the RMF (achieve synchronism). But at the moment, the bar is no more cutting the RMF's flux lines, since it is moving in synchronism with the RMF. So the current flow in the bar will reduce and loses it's own flux lines. Without the flux line 'string' of magnetic force to hold onto the RMF, the bar will fall back from the synchronous speed. But as it falls back, it is again cutting the RMF's flux lines thereby gaining current flow and consequently it's own flux lines. With regained 'strings' of flux lines to hook onto the RMF, the bar will again try to catch up with the RMF. This continuous catching up and falling back all throughout the rotation of the bar around the motor stator in effect causes the rotor to be moving slower than the RMF by about 5%. This 5% is called the slip. The synchronous speed in RPM (Revolution Per Minute) is given by:

$$\eta_s = 120 \qquad\qquad (22)$$

Thus an induction motor moves at:

$$\eta_s = (120) \ X \ 95\% \qquad\qquad (22)$$

Synchronous motors turns at synchronous speed which is the same speed as the RMF in it's stator, generated by say a 1000 miles away generator's rotor. An induction on the other hand turns at 5% less than that speed.

The P in the equation is the number of poles of the motor. For the two pole delta connected induction motor shown in Fig. 37, each phase is joined to a coil and the other end of the coil is joined to another phase. Thus each phase forms only one coil which is a solenoid having one N and one S. One phase wire forming one N and one S which is two poles.

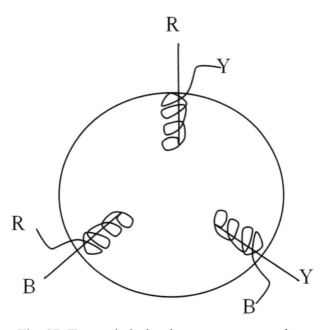

Fig. 37: Two pole induction motor connection

The most common induction motor is the four pole motor. In this motor each phase is connected as shown in Fig. 38; each phase forming two coils or two solenoids. With two solenoids there will be a N and S for one solenoid and another N and S for the other solenoid, that is four poles. Thus the number of N and S poles formed by each phase wire is how the number of poles of a motor is categorized. The four pole motor speed is 1800 RPM (1500 RPM). This is derived as 120 X 60/4 = 1800RPM (120X50/4=1500 British). In a standard four pole motor, each of the phase will form two coils so there are six coils.

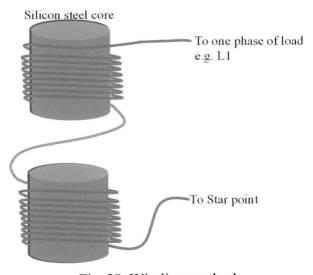

Fig. 38: Winding method

If each phase wire forms a coil and the other end continues to form another coil, four poles are formed. N and S on clockwise coil and S and N on the anticlockwise coil. This is a four pole motor which will spin at 1800 RPM (using 60 Hz) and 1500 RPM (using 50 Hz). This is the most common motor in use.

Now if we look at the squirrel cage, when induced current is maximum on one particular bar, it will be of reducing strength in the nearby bars but on all these nearby bars, current will flow in one direction (say from left to right). This induced current flows to the end rings (left to right in Fig. 39) and then flows down the right end ring, continue travelling in the bottom squirrel cage bars from right to left, reach the left end ring and so forth. Therefore a circulating current is formed which flows from left to right in the top bars, down to the bottom the left end ring, continue flowing from right to left at the bottom bars, into the left end ring and back up to the top of this end ring and repeat itself. This circulating current creates magnetic flus lines that interact with the stator flux lines produced by the generating station.

Squirrel cage of an induction motor

Fig. 39: Squirrel cage of an induction motor

Induction motors have a problem in that it spins at a fixed speed or RPM (Revolutions Per Minute) which is synchronous with the generator in the power station minus 5%. This hampered it's wide use previously. It is only after the invention of the IGBT (Insulated Gate Bipolar Transistor) in 1982 that the use of induction motor started becoming widespread. With an IGBT, induction motors can be controlled precisely from a speed of a few RPM to much higher than synchronous speed (power station generator speed). The fastest speed of an induction motor if just connected to the grid is if there are only two poles. The η_s at 60 Hz:

$$\eta_s = (120) \; X \; 95\% \qquad (23)$$
$$\eta_s = (120) \; X \; 0.95$$
$$\eta_s = 1710 \; RPM$$

at 50 Hz:

$$\eta_s = (120) \; X \; 95\% \qquad (24)$$
$$s = (120 \tfrac{50}{4}) \; X \; 0.95$$
$$\eta_s = 1425 \; RPM$$

In an induction motor, the rotor is made of bars of conductors (aluminum or copper) arranged in a circle and held together by conducting rings at both ends. In effect it is shaped like a squirrel cage, thus the name squirrel cage induction motor. The squirrel cage bars are arranged slightly skewed and not parallel to the shaft to:

1. reduce humming noise
2. to avoid stalling the motor

Squirrel cage bars and end rings made of aluminum will have low mass and therefore a lower centrifugal force is needed to turn it (easier starting). Squirrel cage bars and end rings made of copper can start with higher loads therefore have higher overall efficiency and thereby can reduce of CO_2 emission. Within this squirrel cage is stacked laminated steel sheets of steel for the following reasons:

1. Steel (about 96% iron) will allow better flow of magnetic field lines (better permeability) than free air. Iron is the best conductor of the magnetic flux lines. Thus the magnetic flux lines can be stronger around the squirrel cage bars as current is induced and flow within them.
2. A squirrel cage structure by itself is not capable of carrying a big load. Imagine an empty squirrel cage structure carrying a load requiring 300 HP for example; even it's look doesn't depict a lot of mechanical strength. The laminates of steel provide mechanical strength to the squirrel cage structure.
3. Laminates of steel with varnish in-between will prevent the formation of huge eddy currents (storm of electron flow) which can affect the designed interaction of the circulating stator flux lines with the squirrel cage bar's flux lines of the rotor. This eddy current, if unchecked can also produce lots of heat in the rotor. Eddy is still formed in the outside body of the induction motor but steel laminates in the stator (around which the stator windings are wound, enable keeping the main steel body of the induction motor a distance away; thereby reducing eddy current in the outside body. This author have tried running a 25 HP three phase motor with only the three phase wires going in and removing the green Ground (Earth) wire. The motor turns but is very jerky making a "de de de de" sound and is physically vibrating. By just connecting the Ground (Earth) wire, the sound changed to, "deeeeeeeee" and the vibration stopped and it is perfectly smooth. So the eddy current in the body of the motor, if not leaked to Ground (Earth) via the Ground (Earth) wire will have a drastic affect to the functioning of the induction motor. This means, in factories with lots of induction motors Ground (Earth) is quite critical. In the factory where this author worked, at one time the whole factory tripped every day at about 6pm. Electricians were called in but the root cause cannot be found for some time. So the contractors planted a whole lot of Ground (Earth) rods in the land behind the factory and the problem disappeared. So that factory has a very good Ground (Earth) now. Thinking back this author has an idea of the root cause of the tripping. This is a hard disk platter manufacturing factory. The activity that occurs at around 6pm is visual inspectors (exclusively ladies) taking a sample of the production discs to inspect for defects. They use a table lamp which is of course of high quality made of iron. In high tech factories we cannot use standard plastic table lamps, customer who come for site visits will be nervous about buying our products. In the metal pipe of the lamp is the three core wire going to the lamp. As the lamp is moved, eventually sharp edges will cut the three core wire causing a L-G (L-E) or L-N fault. These ladies will call technicians who are exclusive male and they will cut the damaged point, join the three wires at the same point and insulate all three wires with insulation tape. Then they put insulation tape over all three wires and it looks beautiful enough to impress the lady visual inspectors. This author has done exactly same thing early in his life and have seen a short occur after a few years. The insulation tape simply conducts after a few years. Of course this will not happen if 3M insulation tape is used but they cost 2700% more. Therefore if a normal insulation tape is used, keep the three joins separated by air. This looks ugly but is the best for preventing short circuits. The tape is only to prevent rats in ceilings or humans from getting an electric shock.

4. If the laminates of steel are replaced by solid steel, the magnetizing of only the squirrel cage bars which is needed for the motor to run will not happen efficiently. The whole solid steel will also have induced magnetism running around in disarray.

Motor Starters

The essential functions of motor Starters are to prevent a sudden inrush of voltage and therefore current into the motor coils. The sudden inrush will cause mechanical stress on the bearings and electrical stress on the motor coils. Described in this section are some of the most common methods that are used to start an AC induction motor. These methods offer some benefits in terms of cost.

Reduce-Voltage Starting

When starting voltage must be lowered. The common types of reduced-voltage starters are as follows. This list starts with the most rough starting (meaning the motor will jump a little) to smooth starting.
1. DOL (Direct On Line) – a MCB is used to energize the motor directly.
2. Star-delta
3. Auto-Trans
4. Primary-resistance
5. Soft starter using SCR (Silicon Control Rectifier) – mostly used with DC motors.
6. VFD (Variable Frequency Drive) or VSD (Variable Speed Drive) or Inverters using IGBT

Reason for Motor Starters

By reducing the voltage to about half prevents coils of the motor from burning because of the high ampere drawn initially times a high voltage is too much power for the coils to handle.

Example for the star-delta circuit:
Initially: P = 240V X (High amperes)
Later: P= 415V X (Low amperes)

Star-Delta Starter

Prior to understanding how the Star-Delta motor starter works the electrical contactor must be understood. The Schematic in Fig. 40 is that of the contactor. The exact mechanical arrangement or position of electrical contacts will vary among different brand but the concept of how contactor works is shown in Fig. 40. If there is no electricity in the solenoid, the NC (Normally Closed) contacts are closed. The word Normally means when no electricity is sent in. If electricity is sent to the solenoid, it becomes a magnet and pulls the top iron bar which separates the NC contacts. At the same time it pulls the bottom iron causing the NO contacts to close.

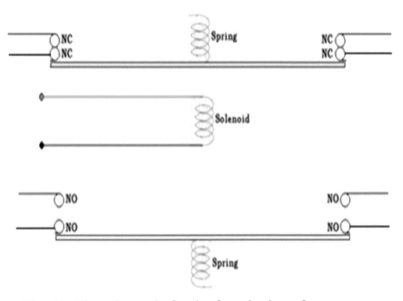

Fig. 40: The schematic for the functioning of a contactor

The power circuit schematic of the Star-Delta is shown in Fig. 41. The contactors in the Star-Delta circuit are named Line (L), Delta (D) and Star (S). Initially L and S close and D is open. Thus looking at the U1, U2 coil of the motor, one side will get 240V from R phase and the other side goes all the way to the S contactor where R,Y,B is looped providing 0V. This will give 240V across the motor coil U1, U2. After 6s, the S contactor opens and D closes giving continuing to give 240V from R phase one side of the motor coil U1, U2 and 240V from Y phase on the other side of the same motor coil. This will provide $240\sqrt{3} = 415$V across U1, U2 coil. The other two motor coils (V1, V2 and W1, W2) will also first experience 240V then 415V.

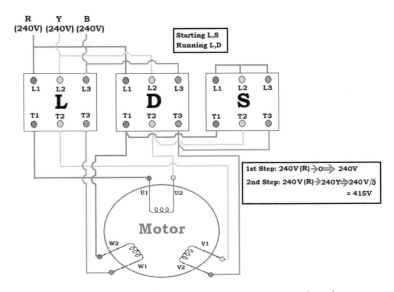

Fig. 41: Star-Delta motor starter power circuit

Fig. 42 below is the control circuit to achieve this. Initially the Start button is pressed allowing energy to go all the way to the coil (solenoid) of S contactor. This will cause the coil to be magnetized closing S1 contact. This will give energy to the three loops whose ends are the coils L, t and G. G is the Green bulb. L coil energized will result in L1 being closed which is a holding circuit for L coil. L2 will hold the push button on which a human finger presses and then releases in a portion of a second. t energized means the central contact in the timer will switch from the NC t2 to open position in six seconds, this will de-energizing S coil. At the same time the NO t1 closes energizing D coil. This will close D1 contact which is a self-holding circuit for D coil. Note the t1 switching is only for a short time similar to the human finger pushing a push button. So a holding circuit is needed to hold the t1 switching. The D1 contact is thus a self-holding for the D coil. The D2, D3 and S2 acts as an interlocking system to prevent wrong sequence of operation of the contactors. This is necessary because sometimes this same circuit is used to operate up to 500 HP motors and wrong operation can be a major disaster.

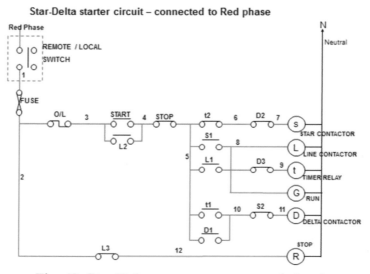

Fig. 42: Star-Delta motor starter control circuit

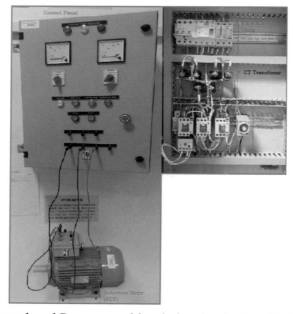

Fig. 43: The Control and Power combined circuit of a Star-Delta motor starter

Autotrans Starter

For the Autotrans circuit in Fig. 44, the names of the three same contactors are changes to L (Line), T (Transformer) and S (Star). The circuit can be understood by looking at one Auto-trans and one motor coil. There are three Auto-trans in one unit of motor starter Auto-trans. Each of these are designed to split each of the three phase voltages L1, L2, L3 (R,Y,B) to 70% initially . That is 70/100 X 120 = 84V (70/100 X 240 =168V). In the second stage (after 6s) the Autotrans is not utilized; the incoming phase wires are directly connected to the motor coils a configuration known as Direct On Line (DOL).

To understand this one Auto-trans and one motor coil is studied. Initially T and S are closed, this will energize the Autotrans. The T will provide L1 (R) phase 240V to the top of the Autotrans coil and the S (above which L1, L2, L3 (R,Y, B) are looped) will provide 0V at the bottom of the Autotrans coil. Thus the wire joint to the 70% point of the coil will provide 70/100 X 120 =84V (70/100 X 240 =168V). This 84V (168V) from L1 (R) phase will go to the one side of a motor coil and another 84V (168V) from Y phase will come from the second Autotrans. With 84V L1 (168V R) phase on one side of the motor coil and 84V L2 (168V Y) phase on the other side of the same motor coil will result on 84√3=145V (168√3 = 290V) across the motor coil.

After 6s, L and S are opened basically taking the Auto-trans from the circuit. And L is closed resulting in 120 V L1 (240V R) phase on one side of the same motor coil and 120V L2 (240Y) phase on the other side of the same coil. This will result in a voltage of 120√3 = 208V (240√3 = 415V) across this motor coil. The control circuit to achieve this is in Fig. 45.

So this motor coil initially experienced 145V (290V) and after 6s it will experience 208V (415V). This is compared with Star-Delta where the motor coil will first experience 120V (240V) and after 6s it will experience 208 (415V). But Autotrans motor controller is smoother for the motor. A 10HP motor can visibly be seen to jump a little with Star-Delta motor starter but not with Autotrans. Primary resistance starters are even more smooth than Autotrans starters. The list of motor starters from least smooth to most smooth are as follows:

1. DOL (Direct On Line) – a MCB is used to energize the motor directly.
2. Star-delta
3. Autotrans
4. Primary-resistance
5. Soft stater using SCR (Silicon Control Rectifier) – mostly used with DC motors.
6. VFD (Variable Frequency Drive) or VSD (Variable Speed Drive) or Inverters using IGBT

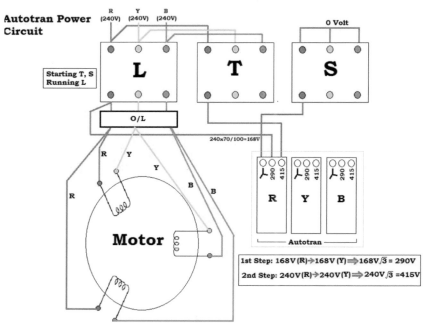

Fig. 44: Power circuit for Auto-trans motor starter

To achieve the above power circuit, the control circuit below is utilized:

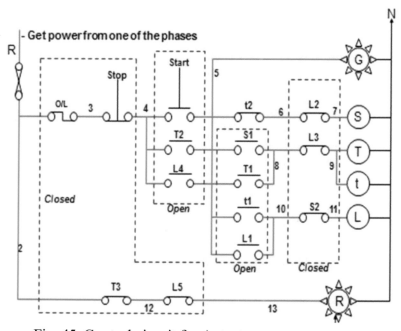

Fig. 45: Control circuit for Auto-trans motor starter

Fig. 46: The Control and Power combined circuit of an Auto- Trans-Delta motor starter

Variable Frequency Drive (VFD)

Currently many motors are being run with VFD but can cost 4000% higher and need an air condition unit to cool it down especially in hot and dusty factory environment. VFDs are also not survivable and so if it fails another 4000% extra cost needs to be spent. By building Star-Delta and Autotrans circuits one can realize a cost effective method of starting motors. The circuit is simple and the parts are easily available and therefore severable. But VFD is only necessary if precise control of the motor is critical to the factory process.

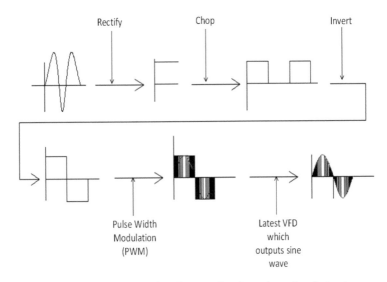

Fig. 47: Variable Frequency Drive (VFD) schematic showing the latest pure sine wave output

Variable Frequency Drives (VFD) or Variable Speed Drives (VSD) or Inverter is used to enable induction motors to have variable speed. It utilizes Insulated Gate Bipolar Transistors or IGBTs) (invented in 1982) as high speed switches to achieve this. This is shown in Fig. 48.

Fig. 48: Insulated Gate Bipolar Transistor (IGBT) used inside a VFD to control a 25 HP motor. The red connector takes in low power signals from a from microprocessor that controls the switching on or off of one of the big 3 Φ phase wires which goes from the screw hole to the induction motor.

IGBT combines the gate drive characteristics of the MOSFET with the high current and low saturation capability of bipolar transistors (BJT). Thus it has an isolated gate FET for the control input and a bipolar transistor as a switch in a single device.

In a VFD line AC voltage at 60 Hz (50 Hz British) is first converted to DC and is chopped to give DC square waves. Six IGBTs are arranged in such a manner that these DC square waves are converted back to AC square waves with a variable frequency. The inverter is sometimes just called motor controller. It enables running the induction motor at precisely controllable speed or torque. The most common VFD chops (switch on and switch off) the pulse such that a thicker pulse represent the high amplitude portion of the sine wave and a thin pulse represent a low amplitude portion of the sine wave. Pure sine wave VFD uses inductors to output a varying voltage pulse at different times thereby following the original power company sine wave almost exactly. Note that in high power applications inductors with taps in-between (Autotrans) are a preferred way to split up voltages just as a voltage divider resistor circuit does in low voltage. Therefore a variable Autotrans in HV replaced the rheostat of LV.

The VFD is also a VSD and an inverter. This inverter term is wrong but is the most common term used worldwide mainly because that is the term used in ever increasing home appliances that uses VFD; including inverter air-conditioning, inverter washing machine and inverter refrigerators. Invert is electrical engineering term which means converting DC to AC, the inverse of rectify which means converting AC to DC. In the inverter schematic shown in Fig. 47 above, converting from DC to AC is only one small portion of the process. Other processes are rectifying and computer control of the IGBT in performing the chopping. Thus a VFD is a rectifier, a computer, a chopper and an inverter combined but is called an 'inverter'. Someone decided to use that term long back and it got stuck just as the wrong term, 'Xerox' is generally used to denote photocopy in the USA. Fig. 47 is an upgrade of the normal inverter called pure sine wave inverter because the final output is a sine wave which is more suitable than the waveform before the last one which is what goes into most inverter controlled motors. Here the middle portion of the square wave has a thicker pulse which has more energy while the sides of the square wave have a thinner pulse which has less energy. This is called Pulse Width Modulation (PWM).

Soon when the price of VFD goes down, it may even appear in home fans where there will be no more 1, 2, 3 speed; a tuner will be able to tune continuously to any speed desirable.

Three IGBTs, one for each of the three Φ wires that goes to the induction motor

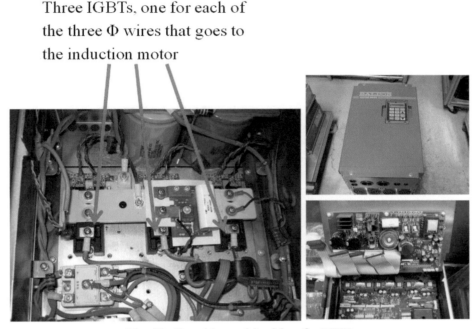

Fig 49: Outside and inside of a VFD

Upon opening up a VFD as in Fig. 49 one can see one or two large capacitors; these are used to produce the DC waveform. VFD causes one defect because chopping waves into square waves is not natural and causes lots of harmonics. Harmonics are many other frequencies of waveforms created other than the fundamental power company frequency. This is especially so at the point of chopping a DC, that is, switching it off. This is at the cliff of the wave. At this point the amplitude of the harmonic can be so high that holes can be poked in the varnish insulation of the motor stator coils. Therefore varnish of VFD controlled motors must be of a higher quality.

In building installations like hospitals (where tiny human body signals are measured) and telecommunication centers (where the bits are getting finer by the day), lots of problems can be caused by these harmonics. So especially in these installations harmonic filters must be installed. A common harmonic filter is an isolation transformer which is a 1:1 isolation transformer. Isolation transformers can cut off 65% of harmonics. The best harmonic filters, called active harmonic filter can cut off 95% of harmonics. Table 3 is the list of available harmonic filters.

Table 3: Various types of harmonic filters in the market

Active Harmonic filters	95%
18 pulse Drive System	95%
Boradband Blocking Filter- Drives	92%
Neutral Blocking Filter	90%
Neutral Cancellation Transformer	90%
Passive Harmonic Filter	85%
Harmonic Mitigating Transformers	85%
AC line reactors	65%
DC reactors for drives	65%
Isolation transformers	65%
Neutral oversizing	

In hospitals and telecommunication centers Grounding (Earthing) must also be as good as possible, that is by using pure copper rods plus cad-welding. The cad-weld as shown in Fig. 50 will ensure a good bonding between the Ground (Earth) cable and the Ground (Earth) rod. Fig. 51 is the black power used as an explosive to achieving the melting of the copper. The black powder itself has a bit of copper in it. In the opinion of this author all Ground (Earth) rods should be cad-welded to the Ground (Earth) wire because the corrosion in-between the Ground (Earth) wire and the Ground (Earth) rod can render the whole system of Grounding (Earthing) useless. Corrosion is copper oxide or iron oxide (cheaper and therefore most common Ground (Earth) rods being made of iron). Considering the price of the long Ground (Earth) green wire and the expensive pure copper Ground (Earth) rods and the manpower to install the whole system, it is worth it to invest in a cad-weld equipment to make a proper join. Corrosion is the formation of metal oxides in this case copper or iron oxides. All metal oxides are insulators.

Fig. 50: Cad-welding of copper strip which is joined to Ground (Earth) rods for a telecommunication tower

Fig. 51: Black power used to do cad welding

Synchronous motor

The basic structure and operation of a synchronous motor is very similar to an induction motor. The main difference in a synchronous motor is that the rotor is not just an un-energized squirrel cage but has coils through which DC is sent to via carbon brush. The workings of a synchronous motor are similar to an induction motor except that the rotor is turning synchronously with the rotating magnetic field (RMF) of the stator. In an induction motor there is a difference in speed of rotor versus RMF called slip. To enable this, the rotor has coils which are energized with DC sent to it via carbon brushes. But at the starting point, the rotor has inertia and cannot attach to the RMF with 'strings' of magnetic flux fast enough. A N pole is initially attracted to a S pole of the RMF will later slip to be in front of a S pole. Thus a rotor coil will experience attractive force to the RMF and then repulsive force. Therefore the motor cannot turn. Thus synchronous motors are not inherently self-starting. To overcome this problem initially the DC to energize the rotor coils is not switched on. A squirrel cage is placed above the rotor coils. This squirrel cage will enable the synchronous motor to be self-starting just like an induction motor. This squirrel cage will carry the coiled rotor along. Once the rotor has reached almost synchronous speed (minus a slip as in an induction motor), the DC to the rotor coils will be turned on. Now the rotor becomes a solenoid and a N pole of the rotor coil will be hooked onto a S pole of the stator coil and the synchronous motor will thenceforth rotate synchronously with the RMF.

The DC goes into the rotor via a pair of slip rings, making the rotor behave like a concentric array of bar magnets. Another option is to use a permanent magnet as the rotor. The need for slip rings and carbon brush are the main disadvantage of the synchronous motor.

All generators are synchronous motors. A recent use of synchronous motors is to replace stepper motors in robot arms. With stepper motors we could mechanically switch the energizing of coils in simple steps as shown in the stepper motor of Fig. 52. In step one, the top coil is energized with to be N then the side coil is energized to be N then the top coil is again energized with a S and then the side coil is energized as a S. With this capability robot arms were actuated almost solely with stepper motors. But today a software called DSP (Digital Signal Processing) which was developed by the telecommunications industry is being used to run synchronous motors with lots of stator coils. Such a synchronous motor can be utilized to move robot arms. The more coils in the rotor, the more precise will be the robot arm's movement capability. DSP software can instruct the energizing of coils precisely so synchronous motors can be used. The advantage of synchronous motors is that they are much more powerful than stepper motors. Most stepper motor actuated robot arms can easily be stopped by a human hand whereas there is actually no limit on the power of a synchronous motor. The current largest motor in the world is the 135,000 HP synchronous motor build by ABB for a NASA wind tunnel.

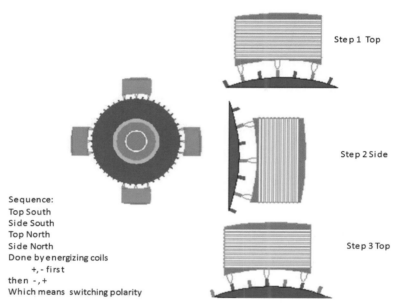

Fig. 52: Stepper motor functioning sequence

Brushless generators

As mentioned previously all generators are synchronous motors. A brushless generator is actually two generators on one shaft. Smaller brushless generator's shaft look like one unit but the two parts are more easily seen on larger brushless generators. The larger of the two sections is the main generator and the smaller one supplies DC current to the rotor of the larger generator. The small generator therefore acts as the exciter mechanism for the larger generator. The smaller generator has a stationary stator coils through which DC current is sent. This will induce current in the rotor coils below it. The consequent AC generated is sent to the shaft in-between the small and big generator. Here the AC is rectified by a rectifier mounted on the shaft. The body of a brushless generator of sizes 20-40 HP has a hole at the point of the shaft where the rectifier is mounted onto the shaft. This way, convection air can cool it down. The resulting DC is sent to the coils of the bigger generator's rotor. Here the coils become electromagnets whose flux lines cuts the coils of the stator above generating AC current. The biggest problem with electronics is the need to cool it down to operate efficiently. Only recently have electronics been invented which can withstands the temperature and dust of a motor shaft, which is why this simple idea is quite new.

Universal Motor

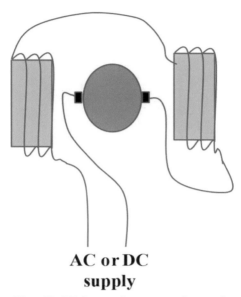

**AC or DC
supply**
Fig. 53: Universal motor schematic

This motor is named universal because it can run on either AC or DC power sources as shown in Fig. 53. Fig. 54 is how the drill universal motor looks like. They have a high horsepower-to-size ratio. They are used mostly as drill or power tool motors. It is also used in vacuum cleaners, food mixers, portable drills, portable power saws, and sewing machines. These motors are rated one HP and below and do not turn at constant speeds, the speed varies with the load. The higher the load the slower it turns which is exactly what is needed when cutting a piece of iron for example. A soft iron piece can be sawed fast with a hacksaw but a hard piece of iron need to be sawed slowly. So when drilling soft concrete, the drill speed is fast and when hard concrete is encountered, the drill bit will automatically slow down. When the motor operates with no load, the speed may attain 15,000 RPM. With a heavy load the motor runs at a few hundred RPM. The rotor of the universal motor is made with laminated iron wound with wire coils just like an induction or synchronous motor. Electric current flows in the stator windings as well as the rotor windings via a carbon brush. The direction of coiling is opposite in the rotor compared to the stator. Say looking at the stator coil from the point of the rotor it is anticlockwise. To get the rotor to have the same magnetic pole at that moment the coiling has to be in the opposite direction (looking at the rotor from the stator). If DC is sent to the stator, it will travel to the rotor coils also. The rotor will have many more solenoids than the stator. At one moment, one rotor coil will be a N, the next a S and the next a N and so forth. The N will be attracted to a S stator pole and the next rotor pole is a S which will repel the stator pole but the first action above has already pushed the rotor enough that this repulsive force enhance the rotational movement. Using AC a positive pulse travels through the stator coils as well as the rotor coil at the same time providing a case as above for the DC case. When the AC switches direction, the rotational movement continues via the attraction and repelling forces.

For electrical contractors it is better to purchase a higher quality drill alike De Walt or Hitachi. There is a very significant difference between a 300 watt drill and a 750 watt one. The difference can be a few seconds per hole versus up to half hour per hole. With the few seconds per hole drill, man-hours and therefore output per worker can be drastically improved. For bad quality drills the outside looks the same but this author have seen the electric motor shaft being held by the plastic casing; without a bearing. For these bad drills, the whole housing of the motor is made of plastic while a Hitachi has the motor housed in a metal housing with good bearings at both ends.

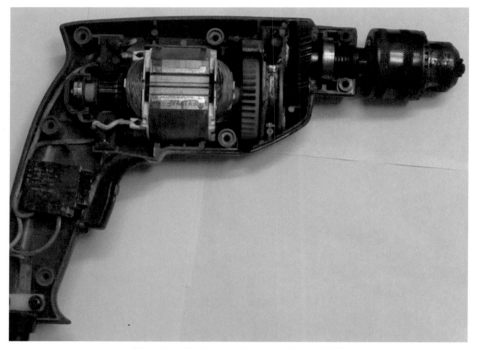

Fig. 54: Universal motor in a drill

Chapter 10
Inductors

Inductors are simply coils of wire which behaves as a magnet as shown in Fig. 55. Inductance is the property of generating an electromotive force (voltage) to oppose a change in current flow. The following story helps in understanding inductors. The young son of this author asked for a piece of copper wire, a nail and a power supply so he could make an electromagnet as depicted in a science experiment book he had just read. This author knew inductors coils always have varnish insulation around the copper wire but the young son said it will not be necessary. What was given to him was one strand of copper wire from a 1.5 mm² cable, a 12 V car battery charger power supply and a three inch nail. The son just coiled the bare Cu wire around the nail and connected the two end terminals to the battery charger. The nail turned into a powerful magnet. But the bare wire did not blow up the fuse in the power supply; though it was much hotter than a solenoid with varnish insulated Cu wire.

Comparatively if the end of a length of uncoiled bare copper wire is connected to the terminals of the battery charger, it is a direct short circuit and the fuse and other component in the battery charger will blow up. As mentioned previously a short circuit can go through protective devices and damage components because a short circuit goes to kilo amperes. But in this case the same bare Cu wire is shaped into a coil and there is no short circuit. Why is happening here? When a bare wire is shaped into a coil, it becomes a solenoid and therefore a magnet. This is therefore a load and there is no short circuit. It is just like placing a permanent magnet above a current carrying wire; the current will slow down under the influence of it's magnetic field and become a load.

Joseph Henry's created his first electromagnet by insulating coiled copper wires with silk strings; today varnish is the standard. The range of varnish is large and the high quality, stable and long lasting motor, generator or transformer is primarily determined by the type of varnish around the copper coils.

It must be noted that some electricians in industry like to coil electrical wires before a termination point; they do this because in future they have this spare length of wire to cut and redo the termination. They can do this but the wire cannot be shaped as a coil. Just remember coils are sensitive to electrical flow. It can be shaped into a long S if needed.

The end of the coil where the electrons flow (opposite of current flow) is in a clockwise direction is the N and the other end is the S of the magnet (somehow the mind can more easily remember and relate clockwise to N so electron flow need to be used). The coil needs to have a layer of insulator normally a thin layer of high quality varnish. The original varnish was from a tiny parasitic lac insect (Laccifer lacca) found on certain trees in India and Thailand. The electrical varnish used today is totally created in labs and manufactured from petroleum and other hydrocarbons. One to four layers of varnish insulation is used. Different films are used for different heat resistance, namely (in order of heat resistance) polyvinyl formal, polyurethane, polyamide, polyester, polyester-polyimide. Electrical engineers give the specification to chemical engineers to develop the varnish and the specification are excellent flexibility (in transformers the laminates in the core need to vibrate), heat shock resistance (transformers and motor coils must withstand sudden high heat), thermal life (transformers and motors need to operate up to 60 years), and abrasion resistance (the laminates in transformers and motors may slide over each other). Most varnish is made to evaporate with the heat of soldering iron to ease joining.

Fig 55: Inductors

Table 4: Charging and discharging of an inductor

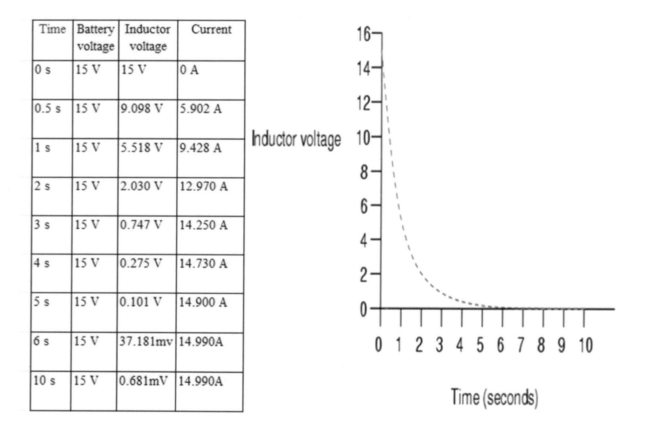

Time	Battery voltage	Inductor voltage	Current
0 s	15 V	15 V	0 A
0.5 s	15 V	9.098 V	5.902 A
1 s	15 V	5.518 V	9.428 A
2 s	15 V	2.030 V	12.970 A
3 s	15 V	0.747 V	14.250 A
4 s	15 V	0.275 V	14.730 A
5 s	15 V	0.101 V	14.900 A
6 s	15 V	37.181mv	14.990A
10 s	15 V	0.681mV	14.990A

As shown in Table 4 inductors store charge for a while before discharging it like a capacitor. In this case it almost completely discharges it magnetic field (which provides voltage) in about eight seconds. After discharging it is more like a short with a voltage of 0.681V (note voltage across a component is proportional to the resistance across it); a piece of wire without a magnetic field to slow current flow so the current shoots up to 14.99 amps.

Inductor's (solenoid) magnetic strength increases with voltage. A physical experiment with a FLUKE 1AC-E II VoltAlert detector indicates this. It only light up and makes sound when it is placed on the point of the circumference of a three core flexible wire (computer incoming wire) where the live wire is. The L wire portion of a three core wire is the only portion of the circumference which has voltage. So only voltage creates magnetic flux lines which triggers it. The difference between the L wire and the N wire is only the voltage; the current must be the same but in opposite direction or else the incoming RCD will trip.

The basic formula for field strength is:

$$B = k_1 NI \qquad (25)$$

Where B is the field strength and N is the number of turns in the coil. K is a constant. The current is given by:

$$V = IR; I = \frac{V}{R} \qquad (26)$$

Where R is the resistance of the wire. This resistance is proportional to the length (L) of the wire. The length is proportional to the number of turns (N) of the wire.

$$I = \frac{V}{R} = k_2 \frac{V}{N} \qquad (27)$$

Putting this back into

$$\text{Yields } B = k_1 N (k_2 \frac{V}{N}) \qquad (28)$$

$$B = k_1 K_2 V \qquad (29)$$

So the field strength is proportional to the voltage. And to increase field strength one need to increase voltage. So multiple turns may not be necessary, a single turn will suffice but a single turn will have too low impedance and therefore too high current which can burn the wire.

When designing a solenoid, start with the current carrying capacity of the wire. Then divide the supply voltage with this current to get the required impedance. From the impedance required (say Z Ω) the number of turns required can be calculated to achieve the required impedance. Say ten turns is first made and the impedance measured with a multimeter (say Y Ω) then the

$$Y\ \Omega \twoheadrightarrow 10\ N \text{ where } N = \text{ one turn} \qquad (30)$$

$$1\Omega \twoheadrightarrow \frac{10N}{y} \qquad (31)$$

$$\text{Therefore } Z\ \Omega \twoheadrightarrow \frac{10N\ X\ Z}{y} \qquad (32)$$

Which is the required turns needed to achieve the impedance required.

As mentioned previously the slowing down of current flow imposed by an inductor is called inductive reactance and follows the equation:

$$X_L = 2\pi fL \qquad (33)$$

Chapter 11

Capacitors

✧

Capacitors are basically two plates with a dielectric in-between them. The formula for a capacitor is:

$$C = \frac{\varepsilon_o \varepsilon_r A}{d} \qquad (34)$$

Where: ε_r = Permittivity of the material

ε_o = Permittivity in free space

A = Area

d = distance between plates

To increase the capacitance value, the A in-between the plates can be increased or the value of d between the places can be reduced. In a high capacitance capacitor, the A is increased by coiling two sheets of aluminum foils with a dielectric in-between which will enable achieving a higher A. The best dielectric is which is found naturally in the ground in certain spots of the earth. With mica as short a distance in-between the plates as possible, thereby increasing C. Mica is the preferred as a dielectric because of it's high dielectric strength and excellent chemical stability. The dielectric, mica enables the A to be large and also the d to be small to achieve high capacitance. Fig. 56 and Fig. 57 shows the outside and inside of a capacitor. Recently the word dielectric strength has become a replacement for insulation strength in most electrical books and journals. This is just like chemical engineers using the word polymers to replace the common word plastic.

Fig. 56: Capacitor

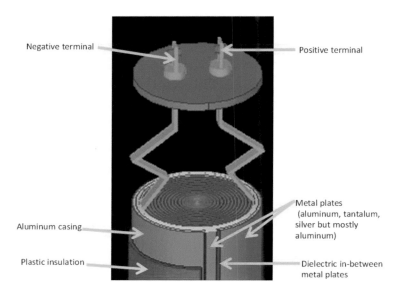

Fig. 57: Capacitor inside, coiling the two metals with dielectric in-between increases A

For a capacitor the formula is:

$$I = C\frac{dV}{dt} \qquad (35)$$

When dV = 0V no current flows
For an inductor the formula is:

$$V = I\frac{dI}{dt} \qquad (36)$$

When dI = 0 no voltage is experienced

Chapter 12
Transformers

Michael Faraday deduced the Faraday's law of induction which is now the fundamental principle of electromagnetism and transformer operation. He invented the transformer in 1831 after inventing the motor and the generator soon after. Faraday wrapped two wires around opposite sides of an iron doughnut and found that if he passed an AC current through one, it immediately induced a current in the other.

Functions of a transformer are:
1. Step-up step-down
2. Isolation – for protection of controllers
3. Impedance matching
4. Remove dc component from ac waves
5. To transfer ac signal without using brushes or other moving contacts
6. Get rid of noise and spikes

However transformers are mostly utilized to raise or lower voltage. It has no moving parts and is simple, rugged, and durable. A transformer is the most efficient machine in the world; low power transformers can have efficiencies of 80-90% while high power line transformer can have efficiencies of up to 98%. This means the energy going out of a transformer over the energy going in, is 98%. Comparatively the energy of a car movement over the energy of the petrol is typically only 25% (maximum of 37%).

A transformer is basically two inductors placed next to each other with an iron core connecting them. It uses the property of electromagnetic induction to get current from one coil to the other without any electrical wire connection. The coil with the incoming electricity is called the Primacy coil and electricity comes out of the Secondary coil.

For the magnetic flux lines to travel from the Primary coil (where current goes into the transformer) to the Secondary coil (where current goes out of the transformer), the best medium is iron. In other words, iron is the best conductor of magnetic flux lines, just as silver is the best conductor of electric current (copper is number two and aluminum is number three).

As an indication of utility of the iron core, this author have tried taking out the iron core from the solenoid of a contactor and energizing it; there is magnetic pull for a screwdriver placed near the coil but it is very weak. Placing the iron core back in the center of the coil turns it to a powerful magnet for the screwdriver placed nearby. What is happening is that magnetic flux lines can travel easily through the iron thereby greatly increasing it's magnetic strength.

Inductance is the property of generating an electromotive force (voltage) to oppose a change in current flow. Though iron conducts electricity, in transformers it is instead used only to conduct the magnetic flux lines from the Primary to the Secondary. Current flowing in the varnish insulated wires of the Primary will generate magnetic flux lines around the Primary coil. This magnetic flux lines are continuously changing direction because AC current is sent to the Primary. A varying AC current creates a varying number of flux lines which is a prerequisite for current induction in the Secondary coil. If DC is sent to the Primary coil, it will be a strong magnet with lots of flux lines but it will not be varying so there will be no current induction in the Secondary coil. But when the DC is first switched on there is a change in number of flux lines from zero to full so a bulb joined to the Secondary will light up once and after that it will never light up even if a powerful DC is flowing in the Primary (resulting in a dense high number of flux lines but no change in them).

There are three losses in transformers, namely eddy current losses, hysteresis losses and copper losses. Transformers are rated in VA or kVA because copper loss (emission of heat from copper coils, $P_{Loss} = I^2R$) depends on current and core loss (emission of heat from the iron core) depends on voltage. Hence the phase shift of V and I or PF (the degree to which the V and I wave are not moving at the same time) has no effect on transformer energy consumption.

A transformer is mainly used for stepping up or stepping down voltage or current following the formula:

1,2, 1,2, 2,1

$$\frac{V_1}{V_2} = \frac{N_1}{N_2} = \frac{I_2}{I_1} = \frac{\sqrt{Z_1}}{\sqrt{Z_2}} \tag{37}$$

$$\frac{V_p}{V_s} = \frac{N_p}{N_s} = \frac{I_S}{I_P} = \frac{\sqrt{Z_p}}{\sqrt{Z_s}} \tag{38}$$

Where N_p is the number of turns of the primary coil and N_s is the number of turns in the secondary coil. V_p and V_s are the primary and secondary voltages, I_p and I_s are the primary current and secondary currents and Z_p and Z_s are the primary and secondary impedances.

In this author's experience of teaching this subject, memory of this formula is an Achilles heel for many students, who tend to forget it during examinations. A slight mistake in memory can cause hugely wrong results. So a three step method was devised to remember the formula. The first step is remembering 1,2,1,2,2,1; like remembering a phone number, then V_1 over V_2 equal to N_1 over N_2 equals to I_2 over I_1 and finally V_p over V_s equal to N_p over N_s equals to I_s over I_p. Perhaps because the words Primary and Secondary are long words, they do not stay in memory easily. It is easier for humans to remember numbers like a phone number.

As mentioned previously transformers are used to step up voltages from the 11-13kV output of generators to 275 kV or 500kV used for transmission. This will reduce the current. Typical current in 275 kV lines range from 10 to 59A; as a comparison a typical car battery outputs 70A upon startup. But it is safe to touch a car battery terminal even during startup because the voltage of the battery is only 12V which is below 40V required for current to jump into a human skin. Note if current does enter the human body, 1A is enough to kill; the 70A at startup of a car does not kill because voltage is 12V.

In 1905 paper, Albert Einstein established relativity which showed that both the electric and magnetic fields were part of the same phenomena viewed from different reference frames.

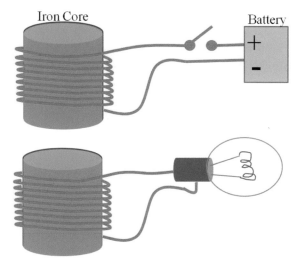

Fig. 58: Transformer logic

Looking at Fig. 58, if the switch is switched on, the bulb will light up once and will not light up again. What is happening is that the magnetic flux lines from the Primary increased from zero to high (a change) resulted in the bulb lighting up once. After that it will not light up again. So to get the bulb to light up again, switch on and off the switch again. Therefore to get the bulb to continuously light up, switch on and off the switch very fast. Alternatively the blue battery can be replaced with AC leads of a home. What is happening in AC is, say the L lead of the home is L1(R) phase. At the power company generator (say about 1000 miles away), the rotor magnet just before crossing the L1(R) phase coil is switched off, as the rotor magnet passes the L1(R) coil, the home AC lead will receive a high portion of the AC wave and when it completes passing the coil, the home AC leads will have a low portion of the AC wave. In effect, the far away generator is a on off switch replacing the fast moving hand to switch on and off the DC supply of the Primary coil.

This switching on and switching off, changes the numbers of magnetic flux lines that are cutting the Secondary coil. The increasing or decreasing number of magnetic flux lines cutting the Secondary coil is necessary for electrical induction. If a DC is sent to the Primary, lots of stable flux lines will travel to the Secondary coil but these flux lines are not changing so there is no electrical induction in the Secondary coil. This can be compared to a magnet being passed a copper wire. Around the magnet are the flux lines as science teachers showed when we were in school (iron filings on a piece of white paper and a permanent magnet placed below). As the magnet approaches the Cu wire the number the flux lines gradually increases (changes) and this result in induced current flowing in the Secondary following Fleming Right Hand Rule which requires F, B and I where F is represented by the thumb, B the pointing finger and I the middle finger. In the transformer's case there is no F of force of movement of the Primary coil but the Primary coil carries AC which creates flux lines which gradually increases goes down to zero and gradually increases again in the opposite direction and so forth. This change in the number of flux lines induces current in the Secondary coil.

The fact that induction of current requires a changing number of flux lines is analogous to $F = ma$ where a changing speed or acceleration (a) is necessary to achieve a force, a constant speed do not result in a force. In a transformer, a changing number of flux lines (equivalent to acceleration) is needed.

The Fleming's Right Hand Rule states that F (force), B (magnetic field) and I (current) must be present to generate current in a coil (in this case Secondary coil). If DC is sent to the Primary, the F is missing therefore there is no I. Actually a bulb connected to the Secondary will light up once and that it will never light up again. The lighting up once is caused by the changing of magnetic flux lines from zero the full as the DC is switched on. After that there is no change in flux and therefore no induced I. With AC in the primary, F is provided in the mechanical F in the generating station far away, thus giving an I in the secondary.

Losses in a transformer

As mentioned previously there are three losses in a transformer, eddy current loss, hysteresis loss and copper loss. A deeper description of these will be given in the next few paragraphs.

Just as any conductor, iron has a problem in that eddy current flows which is like a storm of electricity within the core. Just as a storm or hurricane is created by a variation in temperature and pressure in nearby places, a storms of electron flow is created within the core iron. And any electron flow generates heat and this heat loss is termed eddy current loss. Eddy current in the core iron causes other problems for the transformer because eddy currents will form their own magnetic flux lines which will interact with the magnetic flux of the coils thereby interrupting the transformation function. The resulting AC waves coming out of the transformer is not a clean sine wave and will have distortions caused by the eddy current. The solution to eddy current loss is to use laminates of silicon steel, with varnish in-between. This way the eddy currents can only circulate within one laminate; thereby reducing them drastically.

The next loss is hysteresis loss. Domains are found in all ferromagnetic materials. Iron is used as the base of most magnets, since it is the cheapest ferromagnetic material. Within ferromagnetic material like iron tiny regions called domains exists where unpaired electrons spins align making each domain powerful magnets. But because the unpaired electrons spin in different directions in neighboring domains the overall magnetism in iron is zero. But after swiping a magnet over a piece of iron needle, as the young son of this author did for forty times (following a science guide book), the iron needle became a magnet. What happened in the needle is that the unpaired electron spin direction of all the domains started to be in the same; turning the steel (95% iron) into a magnet. In a transformer, as the power company 60 Hz (50 Hz British) AC is sent to the Primary, the magnetic flux lines changes coinciding with the voltage changes. That is, magnetic flux flows say in one direction to a peak, goes to zero and then builds up in the opposite direction to a peak and goes to zero again. As shown by equation (32), the magnetic flux is dependent only on the voltage variation. The transformer core is silicon steel (95% iron); so this change in voltage causes a change in flux which causes a change in the spin direction of the electron every half cycle. As mentioned above, under the eddy current topic, any flow of electrons will produce heat; this is even more so if the flow is changing direction every half cycle. This loss as heat is called hysteresis loss. The solution is to use silicon steel as the core. Silicon steel means some silicon or sand (SiO_2) is added to the steel to reduce the number of domains within the iron. The silicon must be added carefully into molten iron to form the silicon steel used in the manufacture of transformers. If too much Si is added it will render the iron core incapable of conducting the magnetic flux from the primary coil to the secondary coil (note iron is the best conductor of magnetic flux lines). And adding too little will cause a buildup of heat in the core by hysteresis loss.

One senior contractor in this author's class was asking, "If silicon steel of transformers is just adding sand (SiO_2), why is it sold for such a high price?" The high price is due to the accuracy at which sand must be added to the molten iron.

The basic formula for field strength is:

$$B = k_1 NI \qquad (39)$$

Where B is the field strength and N is the number of turns in the coil. K is a constant. The current is given by:

$$V = IR; I = \frac{V}{R} \qquad (40)$$

Where R is the resistance of the wire. This resistance is proportional to the length (L) of the wire. The length is proportional to the number of turns (N) of the wire.

$$I = \frac{V}{R} = k_2 \frac{V}{N} \qquad (41)$$

Putting this back into $B = k_1 NI$

$$B = k_1 N (k_2 \frac{V}{N}) \qquad (42)$$

Yields

$$B = k_1 K_2 V \qquad (43)$$

This equation shows that the magnetic flux lines created is solely dependent on the voltage. To design a solenoid the following procedure is used:

Start with an initial voltage, say V = 240V, choose current say I = 50A

$$V = IR \qquad (44)$$

$$\frac{V}{R} = R \qquad (45)$$

$$\frac{240}{50} = R$$

$$4.8\Omega = R$$

For 50A a 16mm^2 will which can handle 68A will suffice. Try 10 turns of 16 mm^2 wire plus a 40W incandescent bulb (which measures 106 Ω) as a load. Now there is a coil in series with a 106 Ω load. Now the current in the wire can be measured with a clamp meter. The voltage across the coil can be measured with a voltmeter. With the values of the current and voltage the resistance across the coil can be calculated as:

$$R = \frac{V}{I} \qquad (46)$$

Then by the following calculation the number of coils needed can be calculated as:

$$10T \div \div 0.2\Omega \qquad (47)$$

$$\frac{10}{0.2} \div \div 1\Omega$$

$$\frac{10}{0.2} X 4.8 \div \div 4.8\Omega$$

$$\div 240 Turns$$

Autotrans

An autotran is a single coil transformer with a tap in-between. It has a limit of a 3:1 ratio transformation. So it cannot be used as a RMU (Ring Main Unit) transformer where the voltage transformation is 11kV to 415V.

$$\frac{11kV}{415V} \div \frac{27}{1} or \ a \ 27:1 ratio \qquad (48)$$

Autotrans are used mostly as a motor starter or to change one country's voltage to another. For example: Malaysia (240V) to USA (120V) voltage is:

$$\frac{240V}{120V} \div \frac{2}{1} or \ a \ 2:1 ratio \qquad (49)$$

So to run a USA equipment in Malaysia an Autotrans can be built and if it is a 100 turns Autotrans, at 50 turns the varnish is scrapped off and a tap is soldered to this point.

Malaysia (240V) to China (220V) voltage is:

$$\frac{240V}{220V} \div \frac{1.09}{1} or \ a \ 1.09:1 \ ratio \qquad (50)$$

$$or \ \frac{1}{1.09} \approx 90\% \ or \ Malaysian \ voltage$$

So to run a China equipment in Malaysia an Autotrans can be built and if it is 100 turns, at 90 turns the varnish is scrapped off and a tap is soldered to this point. Of course in both cases the wire size must be chosen in accordance with the amps drawn by the equipment. For example, for a a machine that draws 50A, 16mm² wire can be used because this wire can handle 68A.

Autotrans and tapping of Autotrans are used widely in electrical power systems, most voltage drops to enable the use of advance power electronics like IGBT (insulated Gate Bipolar Transistors) or tyristors uses Autotrans. For example one of the latest IGBT can handle 3.3 kV but how to switch a 11kV housing incoming line? A series circuit of four Autotrans can be built and this will cause a voltage drop of 3.3 X 4 =13.2kV. So a 3.3 KV IGBT switch is placed before each coil. This way each IGBT switch will have to switch only 2.75 kV which is less than it's capacity of 3.3 kV.

Isolation Transformer

Another transformer is the isolation transformer as shown in Fig. 59. These are used to clean noisy AC waves. It is simple a 1:1 turn ratio transformer. That is, if there are 100 turns in the primary coil there will be 100 turns in the secondary coil. Isolation transformers can cut out electrical noise (Fig. 60), spikes or surges (Fig. 61) and harmonics (Fig. 62 & 63) which flow only on the positive side or only at the negative side (meaning DC harmonics). Harmonics are waves other than the fundamental wave that the power company generates. Note even a sine wave travelling only on the positive side is DC (the current does not travel backwards); this is a "dirty" DC wave. Isolation transformers are normal looking transformer but it's ratio is 1:1 as shown in Fig. A0. Isolation transformers can clean up to 65% of the defective waves. To achieve higher percentage cleaning of the waves, the equipment in Table 5 can be used. Filters have become important recently because VFD (Variable Frequency Drives), SMPS (Switch Mode Power Supplies) and CFL (Compact Florescent Lights) causes lots of noise and harmonics.

The load in an induction motor is a balanced three phase load because all three coils of wire utilize the same current or else the motor will be jerking. Since the three phase currents are 120 degrees out of phase, they cancel out each other leaving zero current in the neutral wire. Note the neutral wire is the return path of all three phase wires. In a water pipe system, this return path should be three times the size of each 'phase' pipes. At every moment of time the current in the three phase wires add and minus to result in zero current. So for three phase induction motor, N wire is not connected.

In a normal three phase home however the three phases L1, L2, L3 (R, Y, B) do not use the same current. For an older home in the 1970s or 1980s for example, at one moment of time, the current can be L1(R) = +5A, L2 (Y) = -1A and L3 (B) = -2A. L1 (R) =+5A means current is moving away from the substation to the load with a magnitude of 5A. L2 (Y) = -1A (meaning current is moving away from the load to the substation with a magnitude of 1A). After the loads L1, L2, L3 (R, Y, B) are joined together using the N wire this wire consequently goes to the substation N. The current in the N wire where all the phase wires goes through after the load will be +5A-1A-2A = +2A. This is roughly equal to or less than the current in any of the three phase wires. So the practice since the beginning of AC power usage till today has been to use the same wire size for all three phase as well as N.

But recently three groups of home appliance are creating harmonics which do not cancel each other out, causing higher current at the N wire. The three groups of appliances are motor inverter (VFD in air-conditioners, washing machines and refrigerators). SMPS (Switch Mode Power Supply in computers,

TV, radios, LED lamps and cell phones chargers) and CFL (Compact Florescent Lamp). All these three groups of appliances work differently but have one thing in common; they rectify the power company's AC to DC and then chops the DC into high frequency AC waves. A single VFD motor controller can create up to 25 harmonics. At one moment of time, the magnitude of these harmonics do not add to a low value, making it necessary, as in the USA high-tech centers to specify the N wire to be 50% bigger in size compared to the phase wires. Table 5 is a list of filters to clean out these harmonics. As can be seen isolation transformers clean 65 percent of noise which is cheap and good enough for most installations. This author has heard of installations spending 2400% more than the cost of isolation transformers to clean waves.

The 3rd harmonic is the most concern for power system controllers in factories. The reason is the 3rd harmonic is in phase with the 1st harmonic which is the fundamental power company frequency. Being in phase it can superimpose and get create a higher amplitude waveform. The first few harmonics have high energy and superimposition causes problems but the harmonic number get higher, the energy content is small and superimposition with them do not cause significant problems.

Table 5: Electric power filters available in the market

Active Harmonic filters	95%
18 pulse Drive System	95%
Boradband Blocking Filter- Drives	92%
Neutral Blocking Filter	90%
Neutral Cancellation Transformer	90%
Passive Harmonic Filter	85%
Harmonic Mitigating Transformers	85%
AC line reactors	65%
DC reactors for drives	65%
Isolation transformers	65%
Neutral oversizing	

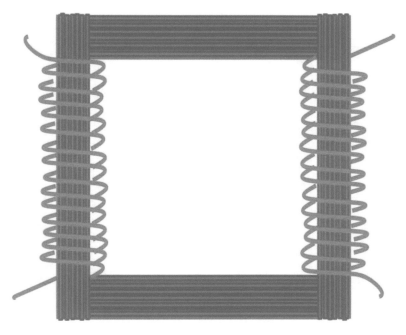

Fig. 59: Isolation transformer same number of turns in primary and secondary

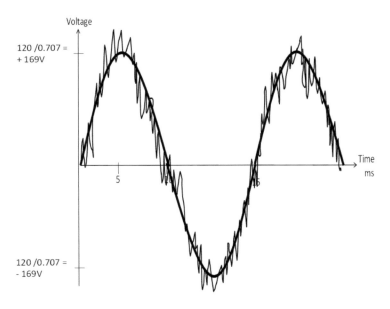

Fig. 60: Noise

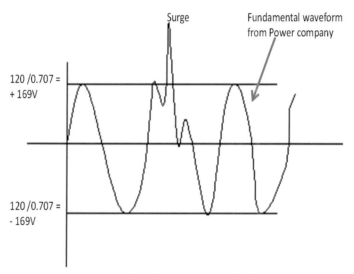

Fig. 61: Surge or Spike

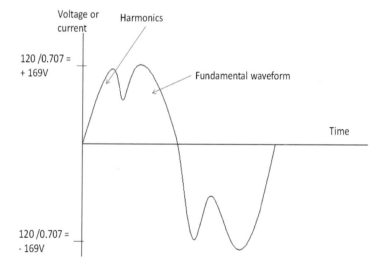

Fig. 62: Harmonics

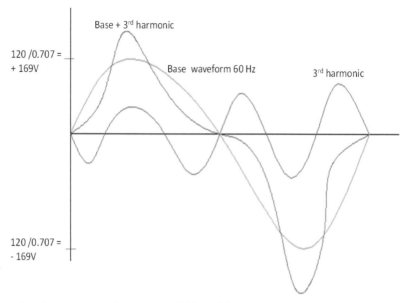

Fig. 63: Harmonics, especially 3rd harmonic must be got rid of

Other transformer developments

Modern scientists are using the transformer principle to replicate Tesla's wireless electric power transmission. At low frequency of 50 to 60 Hz, the Primary and Secondary coils are separated by a paper and an iron core is necessary (for the conduction of the magnetic flux) but if the frequency is increased to very high values like megahertz, electric power can travel through air without an iron core. Scientist from MIT (Massachusetts Institute of Technology) has achieved a coil in front of a classroom transmitting electric power to a classroom of people each having a laptop with a receiving coil. Comparing this with Tesla in 1920s transmitting wireless power to the ionosphere 1000 km away indicates that Tesla's achievement in 1920 is more advanced than the top engineering university in the world today.

The transformer principle is used to operate radio waves. At 60Hz (60Hz) an iron core is needed to transmit AC flux lines from the Primary coil to the Secondary coil. But at 3 kHz to 300 GHz frequencies of radio waves transmission is through the air over long distances. Both Amplitude Modulation (AM) and Frequency Modulation (FM) are used in radio. In AM songs are carried on carrier waves like 690KHz (range is 530kHz to 1700kHz). The 690kHz carrier wave is an AC sine wave and the songs are carried on top of it (multiplexed onto it). A loud portion of the song is represented by a high amplitude on top of the carrier wave and a soft portion of the song is a low amplitude wave on top of the carrier. In FM station like 97.7MHz (range is 88 to 108 MHz) a loud portion of the song is a high frequency wave multiplexed on top of the 97.7MHz carrier wave and a soft portion is lower frequency wave multiplexed on top of the 97.7 MHz carrier wave.

Chapter 13

Electrical formulas

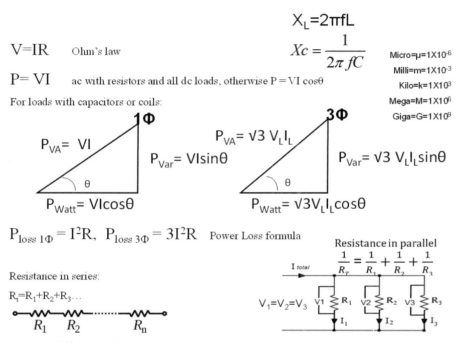

Fig. 64: Commonly used formulas in electrical power

As shown in the middle of Fig. 64, drawing two triangles and putting P_{VA} = VI first then the side of the triangle adjacent to the angle is VIcosθ and the side of the triangle opposite the angle is VIsinθ. For three phase triangle add √3 to the P=VI formula giving P=√3VI but note that the voltage is V=208V and not V=120V (V=415V and not V=240V) in the single phase formula. Then the side adjacent to the angle is P=√3VIcosθ and the side opposite the angle is P=VIsinθ. With this method, six formulas can be remembered using one formula (P=VI for single phase resultant power) and two triangles. A normal human being will find it hard to memorize all six formulas (including this author) without the use of the two triangles. And all these six formulas must be remembered because all are well utilized in the electrical power industry.

PF = cos θ

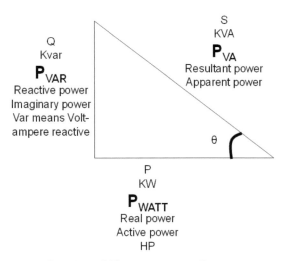

Fig. 65: Different types of power

Fig. 65 shows the power triangle. Various universities and countries use various terms for the three forms of power and they are all represented in the triangle of Fig. 65. The power factor (PF) is the cosine of the angle θ.

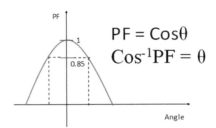

$$PF = Cosθ$$
$$Cos^{-1}PF = θ$$

Fig. 66: Power Factor curve

PF is a measure of how much the current waveform shifts (phase shift) from the voltage waveform. Power companies always want no phase shift at all, so big factories are forced to compensate this phase shift of V versus I with the use of capacitors and PF regulators. PF regulator is the brain that determines how many capacitors to switch on in accordance to the number of coils energized. In the old days not much phase shifting occurred in homes because much of the load was resistive (Edison bulbs, water heaters etc.) but today as humans use more coils as in fans, air-conditioners and florescent tube ballasts, the shifting is quite bad. Also in the old days because most of the loads in homes were resistive, power measuring meter was designed to measure kWh (real power) but as humans changed their lives styles, power companies are increasing losing money because they supply kVA but bill customers kWh. The hypotenuse (kVA) of the power triangle is longer than the adjacent to angle side (kWh). In many countries new meters are installed that measures kVA. In those homes people have experienced their electricity bill going up by up to 30% (hypotenuse is longer than adjacent side by 30%).

The best way to understand the three types of power is by studying a three phase induction motor carrying 5kg load. The work to move up the 5 kg against gravity is the real power in watts. While the motor carry the 5 kg, power is expended to create magnetic flux lines in the stator coils of the motor. This power is called the reactive power in vars. And while this motor is carrying the 5kg mass the power company is supplying resultant power in VA units.

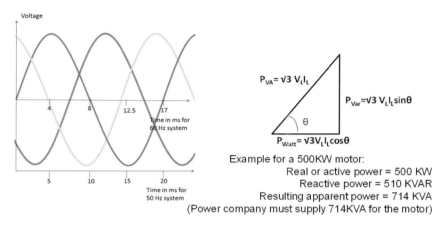

$P_{VA} = \sqrt{3}\, V_L I_L$

$P_{Var} = \sqrt{3}\, V_L I_L \sin\theta$

$P_{Watt} = \sqrt{3} V_L I_L \cos\theta$

Example for a 500KW motor:
Real or active power = 500 KW
Reactive power = 510 KVAR
Resulting apparent power = 714 KVA
(Power company must supply 714KVA for the motor)

‑Work done on a single phase motor is due to the potential difference from phase to N wire (240V to 0V) at that split moment of time

‑Work done on a three phase motor is due to the potential difference between phase to phase wire (415V) at that split moment of time.

‑The active power is converted into useful mechanical power, while the reactive power is needed to maintain the coil's magnetic fields.

Fig. 67: Further detail of motor power consumption

Another real and measured example is the motor depicted by the triangle of Fig. 67. The load carried is 500kW, the coil's magnetic field strength at that moment utilizes a power of 510 kvar and the power company is supplying at that moment 710 kW. It seems logical that if two power is used (one to carry the load and the other to energize the coils), they should be added to get the total power but because the two power are not in the same direction (their vectors are different), Pythagoras theorem should be used to get the resultant power consumption which is the hypotenuse.

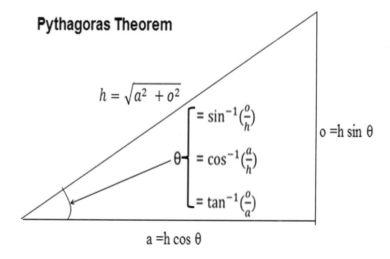

Pythagoras Theorem

$h = \sqrt{a^2 + o^2}$

$\theta \begin{cases} = \sin^{-1}\left(\frac{o}{h}\right) \\ = \cos^{-1}\left(\frac{a}{h}\right) \\ = \tan^{-1}\left(\frac{o}{a}\right) \end{cases}$

$o = h \sin\theta$

$a = h \cos\theta$

Fig. 68: Pythagoras Theorem

The Pythagoras Theorem shown in Fig. 68 is the most utilized formulas in engineering. Civil engineers can be seen on as we drive on roads looking through a theodolite while his colleague holds up a long ruler. He then needs to use Pythagoras theorem and do triangulation to find heights of various spots. This was the main method to find elevation of mountains before satellites and GPS (Global Positioning System) became the norm. Mechanical engineers use Pythagoras theorem to calculate carrying capability of cranes which are shaped like triangles. And electrical engineers use Pythagoras theorem to calculate the three types of power, real, reactive, resultant and angle in-between them (in the form of power factor (PF) which is the cosine of the angle between real and resultant power).

In the navy, if a ship has a friendly ship at say 20 km away which can be seen with a theodolite and an enemy ship is at an unknown distance away but can be seen with a theodolite; a naval personnel will first point the theodolite at the friendly ship and know how far this ship is from his ship. He then uses the theodolite to view the enemy ship location. Now he has the angle in-between the two ships. So he can calculate the distance the enemy ship is from his ship and shoot it with a bomb etc. In Pythagoras theorem, there are three lengths of sides of the triangle and the angle in-between the adjacent (adjacent to the angle) and the hypotenuse. That is four numbers. If one knows any two of these four, one can calculate the rest.

Power Factor

The biggest problem with the usage of AC electric power is the fact that the voltage and current tend to flow at different speeds when moving through an inductor or capacitor; it does not happen when AC flows through resistive loads. Since electric power lines are capacitive (underground cable are more capacitive) and inductive (overhead lines are more inductive) this problem must be rectified especially after the electric power has travelled very long distances. This degree at which they flow at different speed is measured by the angular distance between them, θ and is generally represented as cosine of θ or PF (Power Factor). This process of getting the voltage and current to flow at the same speed is called compensation.

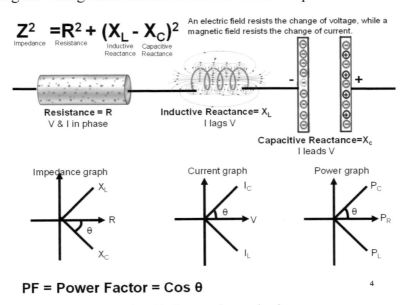

PF = Power Factor = Cos θ

Fig. 69: Power factor logic

Fig. 69 shows electrons within copper colliding with the ions (atoms with less electrons since conductors love to give off electrons and insulators love to take in extra electrons) thereby slowing them down; this is termed resistance (R). The property of conductor atoms to release electrons results in the existence of free electrons within them. If a voltage (Electromotive Force = EMF) is applied to the two ends of a piece of wire, the free electrons will flow as current. Note the electrons do not flow from one end to the other even in DC. An electron will collide with the next one and this will hit the next one and so forth till the end of the wire and this is termed current. The higher the voltage, the higher the current flow and thereby the higher the power as P=VI. Note that current flow (flow of holes) is in the opposite direction of electron flow and there can be no current flow without electron flow. For an insulator, even if a high EMF is applied on two ends of the wire, the electrons are not free to move so there is no current flow. The more the ions within the conductor vibrate (like when it is heated) the slower electrons move through it. This is a general idea that many people have not realized, if lots of insulation is put around conductors as in an underground cable, the hotter the conductor will be. Try being in an enclosed room without air-conditioning. Actually underground cable has another problem because it's structure is that of a capacitor so in addition to resistance due to ions vibrating because of heat, capacitive reactance slows it further.

The old wiring in homes was wires clipped on walls. This was the cool for the wires. These days the wires are run through plastic conduits thereby increasing the heat around conductors. The latest idea is even worse; burying the wires with conduits within walls, making them even hotter. As an indication of the current carrying capacity of plastic covered cables versus bare cables the following statistics are useful:

1. The longest undersea cable (heavily covered by plastic) so far is **580 km, 450kV**, 700MW XLPE HVDC cable between Norway and Netherlands.
2. The longest bare overhead line is the **2,385 km, 600kV**, 7.1GW HVDC Rio Madeira transmission link in Brazil.
3. A close second is the **2,090 km, 800kV**, 7.2GW HVDC Jinping-Sunan transmission link in China.

Therefore cables with insulation surrounding them are far less capable than bare overhead cables. Added to the above statistics, underground cables fail much more often than overhead cables and the overall cost of underground cables is up to 400% higher than bare overhead cables.

In Fig. 67, as soon as a voltage is applied across a coil of wire, a magnetic field builds up around it. There will be a N (when electrons enters it in a clockwise direction) and a S (when electrons enters it in anticlockwise direction). Placing a solenoid below a piece of paper and scattering iron filing on the paper will result in the shape of flux lines exactly as seen when a permanent magnet is placed below the paper. Magnetism is experienced by a coil of wire when electrons are moving in each coil of a solenoid in parallel to each other. In a piece of iron there are domains where unpaired electrons spin around atoms in one direction or parallel to each other. So each domain is a powerful magnet. But electrons around atoms in neighboring domains may have opposite spins so it becomes a powerful magnet aligned in another direction. Thus overall iron is not magnetic. But if a this iron is placed inside a solenoid, all the electron spins in the different domains get aligned in one direction resulting in a powerful magnet. Certain element added to molten iron automatically aligns the magnetism in the domains resulting in a permanent magnet. Otherwise a permanent magnet can be swiped over a piece of steel needle (about 97% iron) about 40 times (as the son of this author did following a science book), the electrons in all the atoms within all the domains start spinning in the same direction turning the steel needle into a magnet.

So as a coil or inductor is energized it becomes a magnet which will slow current flow, just as placing a permanent magnet near a current carrying wire will slow current flow. But it must be observed that voltage creates magnetic field as was shown earlier in this book and the equation:

$$B = k_1 K_2 V \qquad (51)$$

This shows that a magnetic field is dependent only on the voltage. So when the voltage is first applied, a magnetic field is immediately created around the coil. Then slowly the current flows into it. In the electric industry this is termed, current lags the voltage. **Thus current lags the voltage in an inductor.** Eventually after a buildup of magnetic field around the solenoid, the magnetic field will break down leaving zero voltage across the inductor and the process of building magnetic field repeats itself.

Fig. 67 shows a capacitor. Initially both plates of the capacitor are not charged. At this moment, if voltmeter probes are applied across the two capacitor plates, the voltage reading will be zero since there is no difference in the two plates. But when a current is applied to one plate, it gets filled with electrons. Now there is a difference in the two plates. As voltmeter probes are again placed on the two plates, there will be a voltage reading. This is because one plate is not charged but the other is charged with recently filled electrons. Since the electrons must enter before there is a voltage, in the electrical industry this is termed, **in a capacitor current is lead the voltage.** Eventually the buildup of electrons increases to such an extent that there will be an avalanche of electrons to the other plate, and the process repeats itself. Most electrical engineers remember the leading or lagging using capacitors rather than inductors; after remembering how the capacitor works, just remember the inductor has the opposite effect. Some Universities teach students using mnemonics like CIVIL. Going from left to right CIV represents in Capacitor, I leads the V. VIL represents V leads the I in inductors. Some ask why inductor is represented by I. It is simple, current in already represented by I so L is used to represent inductor. Also L is the first letter of the surname of Heinrich Lenz and L is used to honor his work in electromagnetism. In the opinion of this author using mnemonics is dangerous because of the tendency to translate the alphabets wrongly. Remembering how the capacitor works is the best way to remember which one is leading and which is lagging. Current or electrons has to go in one plate before there is a voltage difference between the two places thus current leads the voltage. Note the voltage is always on the X-axis and current on the Y-axis. The current line is always drawn above (leading) or below (lagging) the voltage axis. It can be taken like an anticlockwise race. If I line is above the horizontal V, I has won the anticlockwise race, thus I leads the V.

In industries, capacitors are used to balance the inductive machines which do all the work (mostly motors which have lots of coils). These capacitors are placed in the capacitor bank of the MSB (Main Switch Board) as showed in Fig. 72. Inductive machines like motors, ballast of florescent tubes or induction cookers causes the current to lag the voltage. Capacitor bank units automatically energize the appropriate number of capacitors according to inductive loads used. This switching on of capacitors will cause the current to lead the voltage. So the lagging of current caused by the motor is compensated by the leading of current caused by the capacitors. The 'brain' that switches on or off capacitors is the Power Factor Regulator as shown in Fig. 71. Fig. 73 shows a closer view of the capacitor used in MSB.

Table 6: Effects of PF on a 90KW motor

Power Factor	1	0.9	0.8	0.7	0.6	0.5
Load (kW)	90	90	90	90	90	90
Current (A)	187	208	234	268	312	375

Power company
meter measures this

Current drawn
by house

Power company
meter measures this

Current drawn
by house

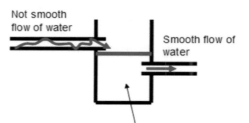

This water tank
is equivalent to
a capacitor

Fig. 70: Capacitor effects

PF Regulator in MSB

Fig. 71: PF regulator

Fig. 72: Capacitor bank

Fig. 73: Typical capacitors this one is rated 109.6 µF and 17.8 kvar at 415V

The schematic for the PF regulator controlling the capacitors is as follows:

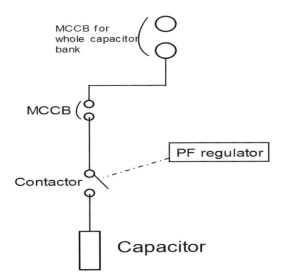

Fig. 74: Capacitor bank single line diagram

As mentioned previously PF is the cosine of the angle between the I waveform and the V waveform. In industry the I wave and the V wave starts to flow at different speed if the load is a motor (a coil) or the ballast of a florescent tube (also a coil). Two methods used to bring back the V and I to the same speed is by use of a capacitor bank or via playing around (slightly) with the DC going into the rotor of a generator (the field coils). Power companies call this running the generator in synchro mode, that is running the generator as a synchronous motor. This is always done in a hydroelectric power station because the hydroelectric generator can easily run as a generator or motor since it is not fastened to any prime mover as in a gas turbine generator, ICE (Internal Combustion Engine) generator or coal fired steam engine generator. A recent third method of controlling the PF is by using a SVC (Static Var Compensator). This is mostly utilized by power companies or large steel mills factories. The SVR has industrial size capacitors and reactors (coils or inductors shown in Fig.104) plus a microprocessor to dynamically balance the phase shift of the I with respect to V. Reactors or any coils have the property of getting rid of harmonics so SVC also has the function of removing harmonics from the power system.

Phase voltage is the voltage between any phase and neutral; while line voltage is the voltage between two phases. In overhead lines the schematic in Fig. 75.

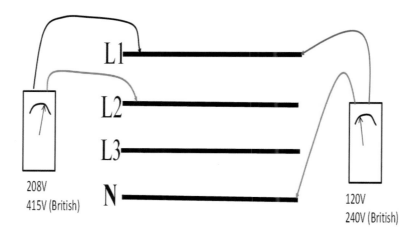

Fig. 75: Single phase and three phase: the standard configuration of housing estate overhead lines. L1,L2,L3, S/W,N (R,Y,B,S/W,N) from top to down.

Chapter 14

Calculations

Sample calculations are as follows:

1. What is the power loss when phase voltage is connected to a 150 Ω resistor?
Answer:

$$V = IR \qquad (52)$$

$$240V = I(150)$$

$$\frac{240}{150} = I$$

$$1.6A = I$$

$$P_{loss} = I^2 R \qquad (53)$$

$$P_{loss} = 1.6^2 (150)$$

$$P_{loss} = 384 Watt$$

2. What is the total resistance of a 20 Ω a 50 Ω and a 100Ω resistor in parallel.

$$\frac{1}{R_T} = \frac{1}{R_1} + \frac{1}{R_2} + \frac{1}{R_3} \qquad (54)$$

$$\frac{1}{R_T} = \frac{1}{20} + \frac{1}{50} + \frac{1}{100} = 12.5\Omega$$

In a calculator just type the following to get the answer:

$$1 \div 20 + 1 \div 50 + 1 \div 100 = 1 \div ANS$$

Where ANS is the calculator key for the previous answer.

3. What is the current flowing through a 30mH inductor connected to phase supply?

$$X_L = 2\pi fL \qquad (55)$$

$$= 2\pi (50Hz)(30mH)$$

$$= 7.42\Omega$$

$$V = IR \qquad\qquad (56)$$
$$V = IX_L \qquad\qquad (57)$$
$$240V = I(7.42)$$
$$\frac{240}{7.42} A = I$$
$$32.3A = I$$

4. What is the current flowing through a 90µF capacitor connected to phase voltage?

$$X_C = \frac{1}{2\pi fC} \qquad\qquad (58)$$
$$= \frac{1}{2\pi(50Hz)(90\mu F)}$$
$$= 35.37\Omega$$

$$V = IR \qquad\qquad (59)$$
$$V = IX_C \qquad\qquad (60)$$
$$240V = I(35.37)$$
$$\frac{240}{7.42} A = I$$
$$6.79A = I$$

5. What is the total power for three factories owned by one owner nearby each other. The first factory uses 50kW at unity power factor. The second factory uses 80kVA at PF=0.6 and the third factory uses 40kVA at PF =0.7 lagging. Also find the total current and KVA.

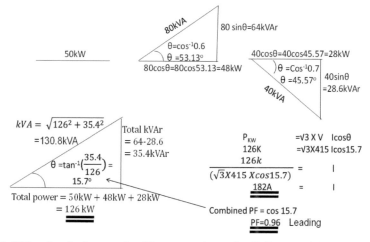

Fig. 76: PF calculation and effects on three buildings with one proprietor

The calculation above depicts a case where three factories are owned by an industrialist. If the power company inspector were to inspect the second 80kVA factory he will report that a heavy fine should be imposed on that factory because the PF = 0.6 while the minimum limit is 0.85 for most countries. Upon inspecting the third 40 kVA factory he will also write up a heavy fine since the PF = 0.7. But if the engineer hired by the industrialist did the above calculation he can show the power company inspector that if all three factories are taken together the PF = 0.96 (way above the minimum limit of 0.85). So there should be no fine for this industrialist.

6. The open circuit voltage of a lead acid battery is 12V. When a 25Ω resistor is connected to the terminals of the battery, the terminal voltage (voltmeter probes on the battery terminals) drops to 10V. What is the internal resistance of the battery?

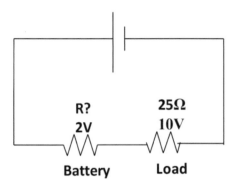

Fig. 77: Battery internal resistance

At the battery internal resistor:

$$V = IR \qquad (61)$$

$$\frac{V}{R} = I$$

$$\frac{2}{R} = I$$

$$Therefore \frac{2}{R} = \frac{10}{25}$$

$$R \times 10 = 2 \times 25$$

$$R = \frac{2 \times 25}{10}$$

$$R = 5\Omega$$

At the load:

$$V = IR \qquad (62)$$

$$\frac{V}{R} = I$$

$$\frac{10}{25} = I$$

7. Given the grouping factor is 0.55, ambient temperature is 0.65, the circuit is protected by a rewritable fuse rating at 15 ampere, calculate the I_z?

I_z=the current carrying capacity of the cable. I_n= fuse rating. The other factors are described below.

$$I_z = \frac{I_n}{\left(C_a + C_g + C_i\right)} \qquad (63)$$

There are five factors that affect the voltage drop namely:

1. Cable table – the standard cable chooser according to the current flowing in the cable. The American Wire Gauge (AWG) or the IEE (Institute of Electrical Engineers) wire current carrying capacity.
2. Voltage drop – taken into account if the load is far away (say 100m) away from the source.
3. C_a – a stands for ambient temperature. If the electric installation is in a boiler room, a bigger wire size need to be used and a freezer project can use a smaller wire size. The C_a can be 0.9 in Saudi Arabia and C_a can be 0.5 in the North Pole.
4. C_g – g stands for grouping. For example this author consulted for a water works project where the architect specified only one conduit for all the cables. So each cable need to be bigger size.
5. C_i – i stands for insulation. There were many types of insulation used previously. One of the first was natural rubber the rubber tree. Today cable insulation worldwide has merged to two types, namely PVC (polyvinyl chloride) and XLPE (cross-linked polyethylene).

So for the above question:

$$I_z = \frac{I_n}{\left(C_a X C_g X C_i\right)}$$

Where In =15A, C_a=0.65, C_g=0.55, C_i= not given so can use 1

$$I_z = \frac{15}{\left(0.65 X 0.55 X 1\right)} = 41.95A$$

Note the value I_z is the current carrying capacity of the cable at these conditions of environmental temperature, grouping and plastic insulation. If the denominator is smaller like C_a=0.1 (as in the North Pole), the current carrying capacity is higher. If the denominator is C_a=0.9 (as in Saudi Arabia) the current carrying capacity reduces. If we put 1, the maximum value, the I_z will be the lowest possible value.

8. A balanced 25A is supplied from 415 V Distribution Board (DB) located 150m away, total voltage drop must not exceed 4% supply voltage:

a) Calculate max permissible V-drop
b) Choose most suitable cable size from table given below:

Cable size (Sq mm)	6	10	16	25
Current rating (A)	34	46	62	80
Voltage drop per Amp per meter (MV)	6.4	3.8	2.4	1.5

The word balance indicates an induction motor which is the only load where the three phases use the same current. The maximum possible voltage drop is 415 X <u>16.6V</u>

The voltage drop equation is:

$$V_D = \frac{(MV)IL}{1000} \qquad (64)$$

We can try 6mm² cable because it can carry 34A which is higher than 25A used by the load.

$$V_D = \frac{(MV)IL}{1000}$$

$$V_D = \frac{(6.4)X25X150}{1000}$$

$$V_D = 24V$$

So 6mm² cable cannot be used because the V_D is higher than 16.6V. 6mm2 cable has a voltage drop of which is above the 4% limit.

Try 10mm2 cable

$$V_D = \frac{(MV)IL}{1000}$$

$$V_D = \frac{(3.8)X25X150}{1000}$$

$$V_D = 14.25V$$

This cable is alright because it's voltage drop is less than 16.6V. The voltage drop is actually, less than the 4% limit. So it can be used for the 150m distance away load.

Chapter 15
Magnetism

In the early days magnetism was thought to be an innate nature of chunks of iron ore struck by lightning. The early Chinese would walk to the top of an iron ore filled mountain to search for iron ore rocks that was struck by lightning; these rocks would be magnetic. They used these rocks to swipe across a non-magnetic iron to make more magnetic iron.

Then in AD1040 Wu Ching Tsung Yao wrote in the 'Compendium of Military Technology' that compass needles could also be made by heating a thin piece of iron, often in the shape of a fish, to a temperature above the Curie Point (a magnet will lose its magnetism if heated above the Curie temperature), then cooling it in line with the earth's magnetic field.

In 1821, Hans Christian Oersted showed that a piece of wire carrying current could also deflect a compass.

In 1820 Andre-Marie Ampere who theorized that magnetism was caused by electric currents and that the same process of magnetism in a permanent magnet occurs in a coil of wire or solenoid.

In 1821, Michael Faraday built the first motor as a piece of DC current carrying wire (positive) hung as a hook on an incoming wire and at the bottom end of this hook was a glass of conductive mercury where the battery's negative terminal was connected. A permanent magnet was placed at the center of the glass. The suspended current carrying wire spun around the permanent magnet, in effect creating the first motor. Today the liquid mercury and the free hook on the other wire are replaced by two bearings holding the shaft of an induction motor.

Later Faraday immediately theorized that the system should be able to work the other way round. He therefore developed the world's first generator where a permanent magnet he moved with his hands generated electricity in a surrounding coil of copper wire. He could measure the voltage of this generated electricity with a voltmeter collected to the ends of the coil.

Thus Faraday was the first to connect electricity and magnetism and so in most literature he is named the Father of Electricity.

In ferromagnetic materials the spin of unpaired electrons in atoms line up in regions named domains. But in neighboring domains the spin of unpaired electrons line up in a different direction thus overall, iron is not magnetic. The magnetic force in each domain of non-magnetic iron is quite strong but it is cancelled out by other domains where unpaired electrons spin in different directions so iron is overall non-magnetic. But if a permanent magnet is swiped over a non-magnetic steel needle 40 times, the needle becomes magnetic because the unpaired electrons in most of the domains start moving in the same direction (clockwise or anticlockwise). An electromagnet can also be used to make the domains line up with one another; thereby creating a permanent magnet. The ability of ferromagnetic materials (iron, neodymium, nickel, cobalt, gadolinium and dysprosium) to conduct magnetic fields is called relative permeability.

When a most of the domains of a ferromagnetic material is magnetized in one direction, it will not relax back to zero magnetization, when the inducing electromagnet is removed. The amount of magnetization it retains is called its remanence (similar word to remember). It is this property that enables the hard disk industry to store information of disks. To drive it back to zero magnetism, a field in the opposite direction must be applied. The amount of reverse driving field required to demagnetize the magnet is called its coercivity as shown in Fig. 78.

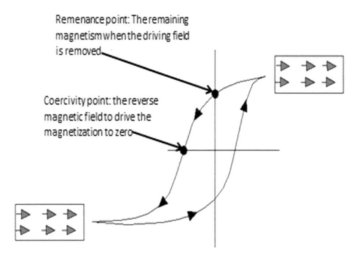

Fig. 78: The Hysteresis Loop

The most powerful permanent magnet uses neodymium. They are used in microphones, professional loudspeakers, in-ear headphones, computer hard disks, wind turbines and hybrid vehicles. 95% of neodymium and other rare earth elements are from China and Hitachi holds more than 600 patents covering Neodymium magnets.

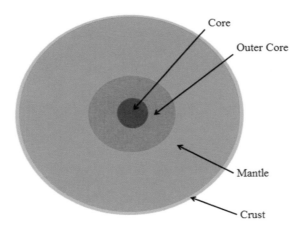

Fig. 79: Magnetism of the earth

The unit used to measure magnetic field is Tesla. The magnetic field strength of the earth is around 0.0001 T. The earth's magnetic field causes a compass placed anywhere on earth to turn such that the north pole of the magnet points to the North Pole of the earth. Note that North Pole is capital letter to indicate that it is the name of a place on earth. Actually North pole is a south pole of the earth magnet which attracts the north pole of all magnets. The magnetism of earth is caused by the molten iron circulating in the outer core of the earth's crust as shown as the second layer from inside in Fig. 79. The inner core is solid iron despite the high temperature due to the extreme high pressure here. The outer core is molten iron and because iron atom loves to throw off two electrons, effectively the outer core is a layer of spinning electrons and these electrons are moving parallel to each other similar to a solenoid; creating a magnetic field effectively making the earth a magnet with the N and S poles. The earth's magnetic field has a great utility in preventing harmful solar and other rays and asteroids from reaching the earth. The reason why iron is found in the inner and outer core is explained like this: If one pours a bit of granular sugar and milk power in a glass and stir it, the heavier granular sugar will converge at the center of the glass; therefore when the earth was just a spinning ball of gas, the heavier iron converged to the center and lighter silicon (sand) appearing on the Mantle and crust above.

The theory of how the moon was formed is that an asteroid crashed onto the earth almost totally destroying earth. In this humongous explosion, a huge cloud of silicon (the Mantle and above is SiO_2) surrounded the earth and eventually this silicon gathered to become the moon. Thus the moon is mostly silicon and therefore has zero magnetic fields around it to prevent harmful rays and asteroids from hammering it daily.

All materials can be classed under three forms of magnetism, namely ferromagnetic, paramagnetic and diamagnetic. Ferromagnetic materials like Fe, Nd, Co, Gd and Dy are the ones strongly attracted to a magnet. Paramagnetic materials: like Al, Mg and O_2 are weakly attracted to a magnet. Diamagnetic materials like C and H_2O and all substances which do not fall in the ferromagnetic or paramagnetic categories.

Chapter 16
Low Voltage wiring

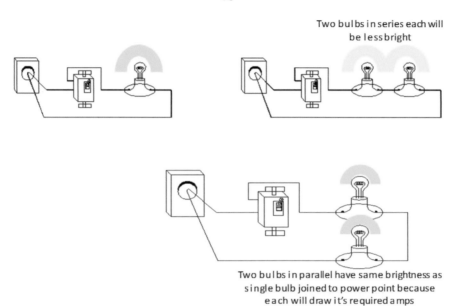

Fig. 80: Home lighting

An experiment was done using two Edison bulbs and the results are in Table 7. Note current is the same in series and voltage is the same in parallel. Fig. 81 is the standard way home wiring is done. Another experiment was done with three Edison bulbs and the results are shown in Table 8. The result in Table 8 and Fig. 82 and Fig. 83 depicts running Edison bulbs without the N wire. Phase to Phase voltage is injected. An interesting observation is that an Edison bulb is able to withstand a large variation in voltage without being blown.

Table 7: Power dissipated in bulbs for different resistance of filament

Incomming supply voltage = 240V	Voltage across each bulb	Resistance of bulb	Power disspated in bulb=V^2/R	Number of times power dissipated in parallel bulb is higher
Series connection of two bulbs	60	106	34	
Parallel connection of two bulbs	120	106	136	4
Series connection of two bulbs	240	160	360	
Parallel connection of two bulbs	120	160	90	4

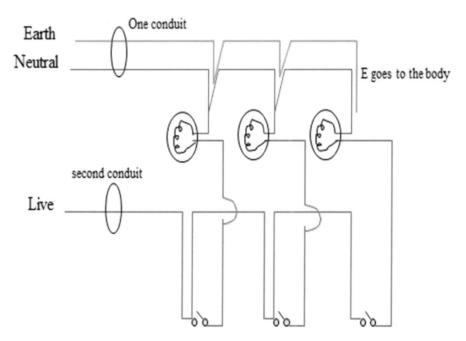

Fig. 81: Home wiring schematic

Table 8: Experiment with 3 Φ voltage

Action	Voltage at terminals of bulb			
	B bulb	Y bulb	R bulb	Bulb parallel to R bulb
RYBN looped on one side of bulb	241	241	237	
N disconnected	247	240	232	
N disconnected and B bulb taken out	360	216	200	
N disconnected and B bulb taken out, R bulb connected in parallel to another bulb	371	292	124	124
YR = 417V				
YB = 415V				
BR = 416V				

Each bulb connected to one phase and N disconnected

Fig. 82: Experiment with 3Φ voltage

N disconnected from supply and one bulb taken out so only two phase supply. Thus this is a series circuit with two phase (R, Y) powering two bulbs but in parallel to one of these two bulbs is another bulb. **Amps determined by single bulb in series which will burn lower wattage equipment.**

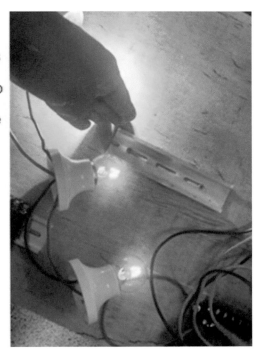

Fig. 83: Further experiments with 3Φ voltage

Chapter 17
Electric Cars

Electric cars will greatly increase electricity consumption over the next few years. Tesla Motors is spearheading electric car development while Google is turning electric cars into self-driving cars. Self driving cars can travel even an inch from each other enabling existing roads to ferry many more cars that it can currently handle. Accidents will be 90% less so hospitals and insurance business will be affected. Overall it is a disruptive technology which even Warren Buffett (considered the most successful investor of the 20th century) is concerned about because he has investment in the businesses which driverless cars will disrupt.

Specifications of Electric sports car made by Tesla Motors

Performance

- 0-60 (0-97km/h) in less than 3.07 seconds.
- 125 mph top speed (201 KM/h).
- Redline at 13,000 rpm.
- 245 mile range (394 KM)
- Full charge in as short as 3 1/2 hours.
- Zero emissions.

Battery voltage: 375 volts

Motor

- 3-phase, 4-pole electric motor, 248hp peak (185kW), redline 13,000 rpm, regenerative engine braking.
- 100% electric motor.
- No camshafts.
- No engine block.
- No turbocharger.
- No supercharger.
- No lubrication system.
- No radiator.

In 2013 the largest electric car sales in U.S. was by Nissan followed by GM, Toyota and then Tesla. In 2015 Tesla took the top position despite it's much higher selling price for cars. Tesla's cars are a revolution in incorporating in the cars the very latest computer and cell phone technologies.

The most expensive portion of an electrical car is it's battery but as with laptop batteries the energy capacity is going up and price is going down. There is also developments being made in induction motor technology such as using High Temperature Superconducting (HTS) cables forming the coils of the motor. The most powerfull magnet in the world are made with HTS coils. The space in-between the rotor and the stator is also made a vacuum to get higher efficiency. From Nikola Tesla's first induction motor till today the main improvement is the reducing the space in-between the rotor and stator. Engineers found that this increased the power of the motor. But if in addition, this space is made a vacuum, the energy density of the motor is increased. The only drawback is the need for a cooling system to cool the coils down to about $90°K$ ($-183°C$). But if that problem can be solved the HTS motors can be about two times smaller and therefore two times ligher. Thus when utilized for electric cars they will be 200% more efficient. Of course it seems far fetched now but we got to know that major developments are not just battries but also in the motor. In fact there is a rush by scientist to achieve room temperature superconductivity. One U.S. navy ship is currently being run with a HTS motor.

Battery developments in electric cars is not just battery chemistry but according to a top person in a high-end German battery manufacturer, the intelligent utilization and coolling of the thousand of battery cells placed at the bottom of the cars is much more important. That person stated that battery chemistry is known by all but the cooling system is now the industrial secret. For example, on one day of driving, a certain sets of battery cells are used and on another day another set of cells are used and so forth.

Chapter 18
Superconductivity

Superconductivity is zero electrical resistance and zero magnetic fields occurring when certain materials are cooled below a certain temperature. This property was discovered in 1911. The ions (atoms with less than normal electrons since conductors love to give off electrons) within the conductors just stop vibrating so electrons can glide over them without any resistance like a MAGLEV (Magnetically Levitated) train travelling over a magnetic field.

For copper or silver there is some electrical resistance even near 0 K. But in superconductors the resistance drops abruptly to zero below the critical temperature.

Superconductive materials like mercury needed to be cooled to near 0K to achieve superconductivity. Then in 1986 a material was created to have superconductivity at 90K (-183°C). So such materials are called high-temperature superconductors (HTS). The latest superconductor can work in 190K but at high pressure. Because liquid nitrogen boils at 77K (-170°C) many applications are now possible with HTS, including:

- MRI machines for hospitals (the most common)
- NMR machines for mass spectroscopy
- Mass spectrometers
- Beam-steering magnets used in particle accelerators
- Magnetic separation; to separate weakly magnetic particles from less or non-magnetic particles (in pigment industries)
- Motors
- Magnetic levitation devices (maglev trains)
- Future uses of superconductors includes:
- Electric power transmission
- Transformers
- Fault current limiters
- Superconducting magnetic refrigeration.

Superconductivity is sensitive to moving magnetic fields so applications that use AC like transformers will be more difficult to build that those that works on DC. Superconductors can maintain a current with no voltage and this fact is already in MRI machines where the magnetism is maintained even though the power is switched off. Superconducting magnets are some of the most powerful electromagnets known. Experiments have shown that currents in superconductors can continue flowing for up to 100,000 years.

Chapter 19
Anode and Cathode

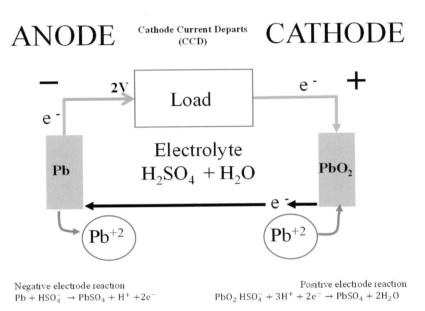

Fig. 84: Pb acid battery Anode and Cathode

Negative electrode reaction
$Pb + HSO_4^- \rightarrow PbSO_4 + H^+ + 2e^-$

Positive electrode reaction
$PbO_2\,HSO_4^- + 3H^+ + 2e^- \rightarrow PbSO_4 + 2H_2O$

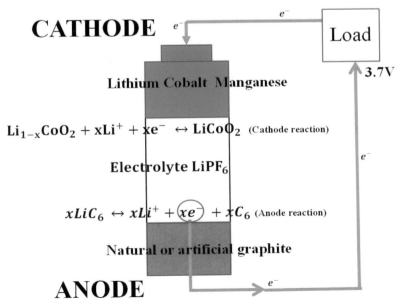

Fig. 85: Li ion battery Anode and Cathode

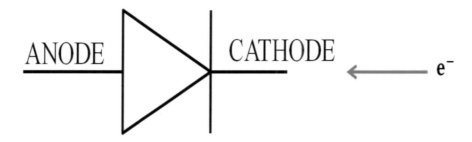

Fig. 86: Diode Anode and Cathode

Fig. 84, Fig. 85 and Fig. 86 depicts the definition of anode and cathode which are words created by Faraday. In a Li ion battery, Li ions move from the anode to the cathode. A cathode is an electrode through which electric current flows out of a polarized electrical device. A mnemonic used to remember this is **CCD (Cathode Current Departs).** Note current flow is in opposite direction of electron flow.

An anode is an electrode through which electric current flows into a polarized electrical device. The mnemonic used to remember this is **ACID (Anode Current Into Device)**. In a galvanic cell (standard battery from a shop), the system has stored energy and if the circuit is completed, the system would like to discharge; electrons (negative) will flow from the negative electrode (which has the "minus" sign) via the wire to the to the positive electrode with the "plus" sign.

Chapter 20

Rectifiers

The original grid of Edison was DC so when it got switched to AC, rectifiers (which convert AC to DC) had to be developed. The simplest rectifier is the transformer bridge diode, capacitor system shown in Fig. 86. It works according to Fig. 87. A diode will cut off the bottom of an AC wave as shown in Fig. 87 top left. A bridge diode circuit provides a full wave rectified waveform as shown in Fig. 87 top right in green wave. But is a capacitor is placed at the output, instead of the green wave going down, it is stored into the capacitor which releases the energy half way down the slope, resulting in a near full DC as shown in Fig. 87 right bottom green wave.

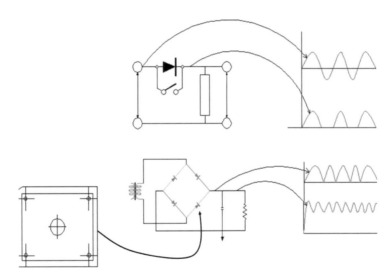

Fig. 87: Full bridge rectifier circuit, box on left is how rectifier currently looks like

Switch Mode Power Supply (SMPS)

The rectifier of Fig. 87 had a DC output that was alright for devices up to the 1980s and 1990s but is too wavy for the latest cell phones, radios, TVs and so forth. So the Switch Mode Power Supply (SMPS) is currently almost universally used as the main power supply for all electronic appliances.

The physical SMPS is shown in Fig. 88 and Fig. 89 is the schematic and Fig. 90 is the waveform sequence. It works on the principle of first rectifying the incoming AC is 120V (240V British) and then chopping it into a high frequency up to 100kHz for the latest cell phones (up to 1MHz have been created in labs). This high frequency 120V AC (240V AC British) is sent to the transformer and stepped downed to say 5 volts for a typical cell phone charger. Then comes the rectifier and capacitor which are used to rectify (the electrical term for converting DC to AC) the 5V AC to a very smooth DC.

As can be seen from Fig. 87 of a typical non-SMPS rectifier; a single diode on the top left schematic just cuts off the bottom half of the AC wave. The resultant wave has a time period of half cycle where there is no voltage. This is enabled by the property of a diode to allow current to flow only in one direction.

On the bottom Fig. 87 is a full wave rectifier where four diodes are used instead of one. This enables voltage to be present over the full time period. The waveform is shown in the Cathode Ray Oscilloscope (CRO) output waveform shown on the top right bottom waveform. This bumpy wave is still DC because it does not go below the 0V line. Note, even a full sine wave above 0V is still DC but a noisy one. To change this bumpy DC into a flatter DC, a capacitor is attached parallel to the output leads. This capacitor is chosen to absorb current over ¼ cycle and release it over the next ¼ cycle. Thus instead of the voltage going down after the ¼ cycle it stays up and goes down only slightly. The resulting waveform is shown on the bottom right CRO output.

The reason the SMPS 5V DC is smooth can be observed if one looks at the CRO (Cathode Ray Oscilloscope) waveform on the bottom right of Fig. 87. The normal rectifier of a power company 60 Hz (50 Hz British) AC wave has the DC going down before being pushed up by the stored energy in the capacitor. But if the frequency is 100,000 Hz as in the latest cell phone charger SMPS, the waveform can only go down a tiny amount before the capacitor release of energy pushes it back up since each wave is packed very close to each other with respect to time. The result is a much flatter DC waveform. In labs 1MHz switching of SMPS has been achieved. Of course the flattest DC is from a battery.

Another big advantage of SMPS is the fact that for any transformer the size goes down with increase in frequency. Thus SMPS of even high amp (note wire size only depends on amps) equipment is quite small. This enables saving of copper whose price have gone up tremendously over time. Thus the latest cell phone chargers are small (small transformer) and outputs a very flat DC. As an indication of the size difference caused by using high frequency waveforms; a standard CPU power supply shown in Fig. 88, the label on the outside shows 450 watt and the 3.3V, 5V and 12V transformers (green top) are rated 28A, 40A and 20A respectively. Their sizes are half inch cube and smaller. Comparatively while this author was building a project at university in U.S. in the 1990s, a 6A, 12V output normal transformer was used, measuring about 7" cube. The weight of that single transformer was much heavier than the whole computer supply of Fig. 88. So if this computer used the same old technology transformer, the transformer itself should be 7" X 20A/6A ≈ 23" cube!

Fig. 88: Switch Mode Power Supply of computer

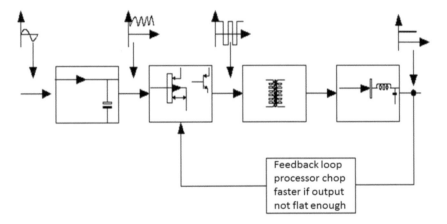

Fig. 89: Switch Mode Power Supply block diagram

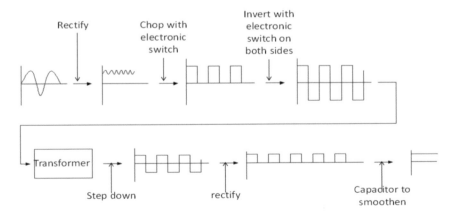

Fig. 90: Switch Mode Power Supply waveform

Detecting rectifier terminals

Detecting rectifier terminals is a very important. In many industries, specifically aluminum smelting, huge rectifiers are used. They are also used to produce the DC to drive the rotors of generators. Mistakes in detecting their terminals can be very expensive. The procedure for detecting transformer and rectifier terminals are as below:

Detecting rectifier terminals: Fix Red multimeter probe on one terminal, Black multimeter probe touching the other three terminals one by one:
Ohmmeter always deflect = Red probe on + terminal
Ohmmeter always do not deflect= Red probe on - terminal
Ohmmeter sometimes deflect = Red probe on AC terminal
Transformer two terminals have high ohm = join incoming L,N join to these terminals
Transformer two terminals have low ohm = join outgoing to rectifier AC to these terminals

If transformer has 5 terminals, the fifth one is joined to the temperature sensor. It has no continuity to any other transformer terminals. Note the high voltage side of the transformer always has the thinner wire size coils and the low voltage side of the transformer has the thicker wire size coils. There was an expensive mistake where a $10 million transformer was destroyed because the high voltage was connected to the larger hole size and the low voltage was connected to the lower hole size. It should be opposite, large wire for low voltage and thinner wire for high voltage. Else just remember that wire size is always dependent only on amps and not on voltage or power.

The coil wire size of a transformer is designed just like any other wire design. First sum up all the loads in the installation in watt, then use P=VI cos θ (single Φ) or P=√3cosθ (three Φ) to calculate the current. From this current the wire size can be determined from the current level versus wire size table. This table is used to determine the correct wire size which will not burn at that current level.

Special cases include long distance between source and load (V_D calculation need to be done), very hot environment (C_a need to be calculated), Compact space for running wires (C_g need to be calculated), abnormal insulation (C_i need to be calculated). All these calculations were detailed earlier.

Chapter 21
Power Generation

Electric power generation is the main polluter of earth. It produces 32% of global warming gasses. The whole transportation industry (cars, trucks, plans and ships) produce 25% of global warming. So more research and technological development need to be invested in attacking this problem. The current world electric power sources are shown in Table 9 and Fig. 91.

Coal (utilizing steam turbines (ST)) is still the main source of electric power producing 40% of it. 37 % of electricity is produced by gas (using Gas Turbines (GT)) and 9% nuclear (using ST) and 4% is produced by oil (using internal combustion engine (ICE)) another 4% with solar and 2% each from wind and others like geothermal and other plants.

Many top thinkers in the electric field believe solar is the future because it is so simple. A ST, GT or ICE used to turn a generator in coal, nuclear, gas and oil power stations are quite complicated. ST or GT have fins that must be positioned extremely precisely for them to function efficiently. A slight movement out of the fins can reduce the efficiency of the GT or GT drastically. Bearing have to be continuously monitored to ensure perfect functioning. A mechanical damage or a small splinter going into the tight spaces in-between GT fins can cause havoc or even a powerful explosion.

Comparatively a solar panel takes out all that complication and seems to have done what microprocessors have done to many previously mechanical machines. Of course there is precision and very high-end science in the manufacturing of solar panels but for the operator or maintenance personnel it is just plug-and-play. Solar panels use the very latest pure silicon used by the highest end computer chip manufacturing or fiber optics manufacturing to achieve highest efficiency of converting sunlight into electricity.

Table 9: World electric power by source (TWh/year)

	Coal	Gas	Hydro	Nuclear	Oil	Others	Wind	Solar	Total
Actual GW	8390	4744	3619	2344	879	541	501	105	21123
% of world usage	39.72	22.46	17.13	11.10	4.16	2.56	2.37	0.50	

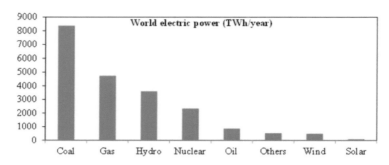

Fig. 91: Electric Power generation sources

Of the total energy produced 66% is generated from hydrocarbon sources. Specifically 40% from coal, 22% from gas and 4% from oil another 4%.

Solar produces 4% and wind and geothermal produces 2%. There is much excitement in the solar power arena because they have reached grid parity, meaning the cost per kW is equal to electricity from coal in some high electricity price regions like California and Hawaii. But solar power is not for all countries. The most successful solar projects are located in deserts or low cloud cover regions. In tropical or equatorial countries where intermittent clouds comes and goes it is not an optimum solution. The clouds in equilateral regions look thin but taking a plane one can observe that it can be as thick as a tall mountain. That will heavily reduce sunlight reaching solar panels.

The problem of solar power utilization in cloudy countries is depicted by the experience of Hawaii in 2015. Dependence on solar power damaged grid equipment because a **cloud cover can take out 70 to 80 percent of solar power output in less than a minute.** Thus the other power stations and switchgears in Hawaii were damaged as the load drew all their demand from the remaining diesel engines and gas turbines. Other than hydroelectricity, no other power stations can load the grid or backup this loss that fast. One of the generators of the Bakun dam in Sarawak can load 300MW to the grid in 20 to 30 seconds). Therefore it is optimum for a tropical and cloudy country to develop both hydro and solar power simultaneously. This author's previous job at a power company was to call the various power stations to start up extra generators or shut them down. When a GT was called to start-up, it would take 45 minutes before the electrical power got onto the grid. For an ICE it would take more than 60 minutes, a coal power station can take eight hours but a hydro power station can load the grid with power in 20 to 30 seconds. Therefore only hydro power station can back-up a sudden cloud cover over a huge solar farm.

Fig. 92 depicts the CO_2 emission by various sources in 2013. The electric power industry produces 32% of the CO_2 emission to the atmosphere. This is a total 6,526 million metric tons of CO_2 emitted to the atmosphere every year and it is increasing every year. Fig. 93 depicts the feed stock for generating electric power currently. Table 10 depicts the top electric power generation countries.

Going by all the above data, the world definitely needs some innovation to counter the effects of global warming which we are witnessing ever more frequently. Examples are the 2012 Hurricane Sandy in NY, USA, 2010 Queensland, Australian floods, the 2014 Philippines flood and the 2014 Kelantan, Malaysia floods are some of the Global warming caused disasters.

Solar power has to be harnessed as much as possible in countries with little cloud cover located mostly in temperate or desert regions. A pure desert which has the least cloud cover has a problem with solar power currently because heat affects solar panel's efficiency. Hopefully innovations can solve this problem.

Justification for Hydro power

Wind power need to be installed in heavy wind areas like the Mid-West of USA and of the coast of Norway and North Sea. In the equatorial regions where rainfall is typically 560% the average level in the U.S., hydropower should be fully utilized. But hydropower requires another factor which is high mountains to enable good gravity drop of the water. For example this author lives in Sarawak, Malaysia which has optimum conditions for hydro power namely:

- Sarawak is the largest state of Malaysia (at 124,450 km²)
- Sarawak has a population of only 2.5 million so large areas can be dammed up.
- Sarawak is located on the equator and therefore has a rainfall of about 4000mm (157 inches) per year compared to a U.S. average of 715mm (28 inches). That is 560% higher rainfall.
- Sarawak has a topography of high mountain ranges at the border with Indonesia.

Brazil is also a rainforest but lacks mountains so the Itaipu dam needed to be built at the border regions with Paraguay. Thus the Itaipu dam's power is shared with Paraguay.

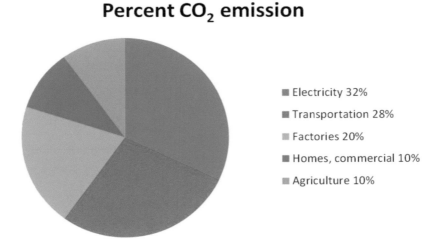

Percent CO$_2$ emission

■ Electricity 32%
■ Transportation 28%
■ Factories 20%
■ Homes, commercial 10%
■ Agriculture 10%

Fig. 92: CO$_2$ emission by various sectors

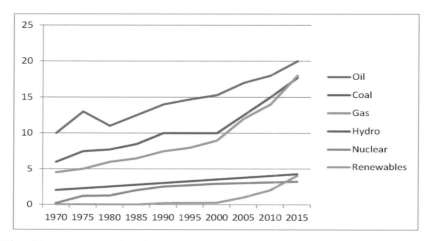

Fig. 93: Power production by source Y-axis is in 1000 TWh per year

Table 10: Electric power production by country

Rank	Country	Electricity consumption (MWh/year)
1	China	6,483,800,000
2	United States	4,686,400,000
3	India	1,111,722,000
4	Russia	1,016,500,000
5	Japan	859,700,000
6	Germany	582,500,000
7	Canada	499,900,000
8	France	462,900,000
9	Brazil	455,800,000
10	South Korea	455,100,000
11	U.K.	323,300,000

Wind power

For wind turbines, GE's advanced technology allows the elimination of gearboxes. Direct drive or gearless turbines reduce the number of moving parts in a unit and increase reliability, helping to minimize costly open-sea maintenance.

Warren Buffer said industrialist will get a lot of tax credit if they build a lot of wind farms which is the only reason he invests in them. Recently Bill Gates explained in an interview with the Financial Times why current renewables are dead-end technologies. He said they are unreliable, battery storage is inadequate and wind and solar output depends on the weather. Bill Gates stated that the cost of reducing CO_2 emission using today's technology in wind and solar are 'astronomical'. Wind power turbines are especially notorious in killing birds. The figure can reach as high as 80 birds per turbine. Therefore the total number of birds killed per year by wind turbines is estimated to be as high as 39 million. But Elon Musk of Tesla Motors recently claimed he has solved the critical battery problem with what he named, 'Powerwall".

Wind power versus the rest

Fossil-fuel plants are cheaper now, but if the UN charter of paying for CO_2 emissions are included, wind farms are cheaper. Also the price of oil can go up drastically in the near future. An equivalent power output deep-water wind turbines and nuclear power station cost about the same. This is also equivalent in price of coal power station with capture and sequestration. Deep water wind farms are only slightly more expensive than onshore wind. Deepwater wind turbines are normally 47 to 70 km off-shore, are normally four legged but there is one unit in the world that floats on the water. There is always more wind the greater the distance from the shore. Some researchers have also envisaged flying balloons with wind turbines mounted on them. The higher off the ground, the higher the wind speed.

Fukushima Nuclear Disaster

Professor Emeritus Kiyoshi Kurokawa from Tokyo University and his team which investigated the Fukushima disaster made a statement that Fukushima cannot be regarded as a natural disaster and it was a profoundly man-made disaster that could and should have been foreseen and prevented. He also said it's effects could have been prevented by a more effective human response to the disaster. He said governments, regulatory authorities and Tokyo Electric Power [TEPCO] lacked the sense of responsibility for the lives of the people and society. He effectively concluded a bit emotionally that the TEPCO effectively betrayed the nation.

Solar energy

Photovoltaic (PV) technology has sprung up recently all over the world greatly aided by cheap solar panels made in China. The idea of using crystals or non-metals in the electrical industry began with Jagadis Chandra Bose in 1895. Bose used it to build a radio receiver. The silicon industry has now grown into the computer/cell phone industry, the fiber optics industry and the PV solar industry.

Elon Musk of Tesla motors stated that it takes only a small square portion of Northern Texas to power the whole U.S. with solar power. And the batteries required are a tiny portion of this small square. He is building a Gigafactory in Nevada to enable this. Of course he is not proposing that all the solar panels be located in Northern Texas though if the U.S. government decides on this route, Northern Texas has an opportune low cloud cover plus cool temperature to enable this; cool temperatures are good for PV efficiency as well as battery life. Elon Musk idea is that the solar panels should be widely sold in individual's homes together with efficient computer controlled batteries he calls the, 'Powerwall'. The solar panels will be on homes and office rooftops so it will not take away crop land. Batteries are the biggest reason for the failure of solar power projects. Even high-end lead-acid (Pb acid) batteries tend to last a maximum of four years. While nickel cadmium (NiCd) batteries can last for 20 years but tend to cost about five times in initial investment. So almost in all projects, lead acid batteries are used by contractors so they can cut cost and collect money for their project fast. But the project is guaranteed to fail in two to four years. Lead acid batteries have been used for a long time in cars where after continual research, they have the lowest internal resistance ($50m\Omega$) but they do not last long. NiCd batteries have a 20 year life-span and has an internal resistance ($70m\Omega$) almost equal to Pb acid batteries. Other major battery chemistry are Li ion ($320m\Omega$) and NiMH ($778m\Omega$), both of which have much high internal resistance The internal resistance is the resistance which causes a battery to discharge even when placed on a shelve because the internal resistance slowly drains the charge from the battery. The only problem is that Cd is banned in Europe for health reasons.

Internal resistance is what causes the terminal voltage of 12V car batteries to be about 14V when the car (or alternator) is running and below 11V if it is disconnected from the car and a load is connected to it's terminals. Car batteries push out up to 70A during start-up. But the battery is only used for start-up, after that the alternator energizes the required electronics and lamps of the car. But if more than 20% of it's charge is used, it starts to die permanently. That is not a problem in cars where battery energy is only used in the cranking and after that the alternator takes over. Pb acid batteries have another problem in that it cannot stand a sudden discharge so it is not suitable to be used as a battery backup in buildings where a power company outage will require a sudden discharge and more than 20% usage of it's stored energy.

NiCd batteries on the other hand do not have all these problems. But they are about four times more expensive. The internal resistance of NiCd is almost equal to that of Pb Acid batteries which is the lowest for common battery chemistries. The most common battery chemistries are Pb acid (voltage per cell = 2V), NiCd (voltage per cell=1.2V), Li ion (lithium ion) (voltage per cell=3.7V) and NiMH (Ni Metal Hydrite; voltage per cell=1.2V). But for a solar project to work NiCd is the only solution. Another problem is that batteries generally survive well only at 25°C and below so in the equatorial and tropical regions or during summers in the Northern and Southern regions air-conditioning for the batteries is a must.

Previously solar panels were made of a series of 12 or 24V arrays. Today 48V system is the norm which results is higher efficiency tapping of solar power due to lower line losses. Solar energy has become feasible option for many poorer countries because manufacturing in China has driven down solar panel cost world-wide by about 62.5% within three years (2012 to 2015). That is, the price have gone from $0.8 per watt to $0.3 per wat).

Another type of solar power is Solar Thermal plants. These power stations are opportune for very hot deserts. The high temperatures can only be better for these types of plants while the high temperature drastically reduce photovoltaics efficiency. In Solar Thermal Plants cheap mirrors (not solar panels) direct sunlight to a central tower to hear up salt to a liquid at 500°C. This hot molten salt in stored in tanks and can boil water to power a steam turbine even at night. This is a huge advantage in that we can finally have non-intermittent solar power. 69 locations world-wide have built such plants with power generating capacity of 4.3 GW. Some boil salt but most just boil water to run steam turbines so they do not work at night.

Tidal Power

Since 2008, a dual-rotor tidal turbine has been feeding up to 1.2MW of power to the electrical grid of Ireland. A 1320MW barrage is being built in South Korea with an expected completion date of 2017. Another 320 MW tidal power plant is being planned in Swansea, UK. As of today, a total of 16 tidal power projects are in various stages of being utilized worldwide.

Fuel Cell

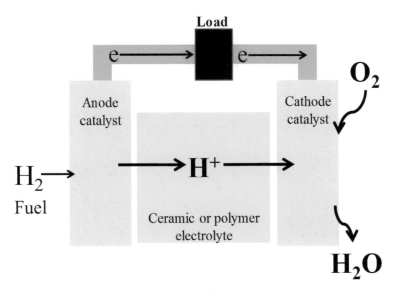

Fig. 94: Fuel cell schematic

At one time fuel cell technology depicted in Fig. 94 was mooted to be the future but it still uses fossil fuels. Basically hydrogen is extracted from natural gas. Hydrogen is the simplest element in the universe with one proton and one electron. At the anode, the proton moves through the electrolyte and the electron powers a load like cell phone. At the cathode, the proton recombines with the electron and oxygen from air to output water. The big problem with this technology is that it still uses fossil fuels plus the energy to take out hydrogen from hydrocarbon is more energy intensive than the output power that emanates from the fuel cell. It is currently well utilized to power servers in Silicon Valley by the likes of Google, Apple, eBay and many others.

Chapter 22
The Air Conditioning system

❧

Air-conditioner is a heat pump whose schematic is shown in Fig. 95. It is a simple but ingenious invention which is very widely used worldwide. The opposite of it is a heat engine which is the combustion engine of all cars, gas turbines and steam engines. A liquefied natural gas (LNG) plant is huge and can reach many miles long but it is basically a huge air-conditioner (air-con) that cools gas to a liquid of -161°C. Heat pump is the engineering term for the air-con cycle (or refrigeration cycle). Basically heat pumps and heat engines are drives human civilization today.

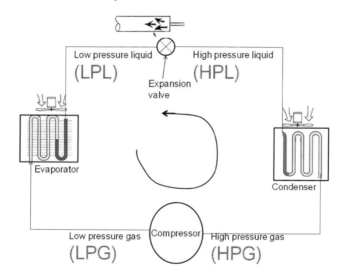

Fig. 95: The air-con cycle

The main formula for the refrigeration cycle is as below. This law is called the ideal gas law:

$$PV = nRT \qquad (63)$$

where

P is the absolute pressure (Unit used: atmospheres, atm)
V is the volume (Unit used: Liter, L)
n is the amount of substance (loosely number of moles of gas)
R is the gas constant, which is the same for all gases
 (Unit and number used: 0.0821 L·atm·mol-1·K-1)
T is the temperature on the absolute temperature scale (Unit used: Kelvin).

The main components of a refrigeration cycle are the cooling or evaporator, the compressor, the condenser and the expansion valve.

As the cold refrigerant liquid moves through the cooling coil in the evaporator, it picks up heat from the air passing over the surface of the coil. When the liquid refrigerant picks up enough heat, it changes to a vapor (gas). The heated vapor is then drawn into the compressor where it is subjected to a higher pressure. As a result of being pressurized, the temperature of the vapor increases.

After coming out of the compressor, the refrigerant vapor is now at a high pressure and high temperature. In this state, vapor passes to the condenser where heat is removed and the vapor changes back to a liquid, however this liquid is still under pressure.

The liquid then flows to the expansion valve, as the liquid passes through the valves, its pressure is reduced immediately. As pressure decreases the temperature of the liquid lowers even more and it is now ready to pick up more heat.

The cool, low pressure liquid next flows into the cooling coil of the evaporator. The pressure in the cooling coil is low enough to allow the refrigerant to boil and vaporize (some refrigerants boil at temperatures as low as -29° C), as it again absorbs heat from the air passing over the coil surface. The cycle then repeats itself as the heated refrigerant vapor is drawn into the compressor.

Chapter 23
The Grid

Transmission Lines

In electric power transmission lines, the Ground (Earth) wire is placed on top or higher than the three phase conductors to protect the phase conductors from lightning strikes. This Ground (Earth) wire is also joined to all the steel poles. Since lightning is only looking for the easiest pathway to reach Ground (Earth) and will always strike the highest point, it will strike the Ground (Earth) wire and will not be interested in the phase wires from which the Ground (Earth) is only found after a load in the substation N. Striking the Ground (Earth) wire, it will continue through the steel pole and drain to Ground (Earth). Normally each steel pole will have four Ground (Earth) rods through which surges can drain to.

Power Regulator Bank (reactors)

Power regulator banks are located along the line usually in a substation. They are utilized to regulate the voltage on the line to prevent under-voltage or over-voltage conditions. They can be manually or automatically switched on via a thyristor. They are basically coils of varnished wire that introduces inductive reactance ($X_L=2\pi fL$) into a circuit. As the grid line is switched into a reactor it acts as a reactance in parallel to the load and therefore reactance reduces (as more resistors are connected in parallel total resistance goes down) so voltage goes down ($V=IX_L$). Therefore as a reactor is switched out total reactance increases and voltage goes up ($V=IX_L$).

Below is an EXCEL calculation of resistor in parallel following the equation:

$$\frac{1}{R_T} = \frac{1}{R_1} + \frac{1}{R_2} + \frac{1}{R_3}\ldots \qquad (64)$$

1/200,000	1/200,000	1/200,000	1/200,000	1/200,000	1/200,000	1/200,000
0.000005	0.000005	0.000005	0.000005	0.000005	0.000005	0.000005

This was done till column KQ which is adding 303 resistors in parallel of 200,000 Ω each. The final resistance of 303 resistors in parallel is 662Ω.

Shunt Reactors

These coils are connected in parallel with the transformer outgoing windings 275kV or 500kV substation. They are used to compensate for the capacitive var reactance of long underground cables.

Defect of AC lines compared to DC lines

- AC require four conductors L1, L2, L3 (R,Y,B,N) compared to only positive and negative for DC.
- AC lines need to be compensated or else the current wave will not move synchronously with the voltage wave (phase shift occurs). This will lead to more current drawn as shown by the power station output equation below:

$$P = \sqrt{3}VI \ cos\theta \qquad (65)$$

If an increasing phase shift occurs, θ increases and $cos\theta$ (or PF) reduces. So to keep P the same value (for example power of water under gravity in a hydro power station is fixed), I has to be increased since V cannot change because the power company keeps it constant. Now putting a high I into the power loss in lines equation below, results in the lines getting hot and wasting energy.

$$P_{loss} = I^2 R \qquad (66)$$

- AC has a skin-effect problem where current only travels on the skin of the conductor and do not utilize the center portion. DC does not have this problem

- AC creates induced currents (power wastage) in the cable sheath and armor.

High Voltage (HV) Overhead lines

Common HV used worldwide are: 1000kV, 500KV, 275 KV, 132 KV, 66kV, 33KV, 22kV, 11KV. Stranded conductors are used, because they are stronger and easier to build (lighter during construction). The center of the conductor is steel strands to provide strength. The center of the cable is not used by the AC current anyway because of the skin-effect where current only flows on the skin of the conductor. The central steel of the cable and the skin effects are depicted by Fig. 96 and Fig, 97 respectively.

Fig. 96: Steel reinforcement in the inner radius of overhead HV
cables (darker color), the outer radius is aluminum

Fig. 97: Skin effect

Fig. 97 indicates the flow of AC in a conductor depicting the skin effect. Current flows mostly in the outer section of the conductor in a region less than a cm thick. The skin depth is termed δ (small letter delta) and increases with frequency of the AC waveform, That is, δ is propotional to the frequency. Cables are run as bundles of two three or four wires together as shown in Fig. 98, this will increase the power carrying capacity of the electric poles. Such cables are officially called bundled cables. The insulators seperating the cables are called spacers and are shown in are shown in Fig. 99. The human hand holding it indicates the actual size.

The skin effect problem in AC cables is due to eddy currents created by the oscillating AC current. It is basically current generation as current moves back and forth over the conductor. The skin effect is caused by the opposing eddy currents induced by the changing magnetic field of AC current. For copper skin depth, $\delta = 8.5$ mm (just under 1 cm); meaning current only flows from the outer surface of the cable to 8.5 mm inside it as shown in Fig. 97. At high frequencies skin depth is smaller.

For 33KV and above cable, the center is made of steel strands as shown in Fig. 96 and the circumference is AAAC (All Aluminum Alloy) $100mm^2$. For 275KV $402mm^2$ is used for HVDC.

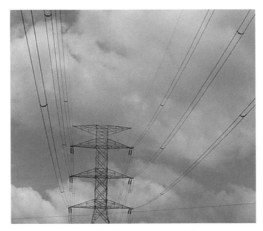

Fig. 98: Bundled cables

Fig. 99: Spaces for bundling cables

When higher amounts of current need to be transmitted, increasing the cross sectional surface area of the cable is not the answer because of the skin effect. The reactance decreases only slowly with size increase. But size increase results in much higher cost and weight; it is harder to install one high diameter cable than many smaller diameter cables. A large size cable carrying very high current will also have corona loss (radiating purple light). It is because of this that cables are bundled as two, three, four or six cables with spacers. Spacers insulators must be built strong to overcome the force of heavy wind and magnetic pull during a short circuit.

Chapter 24
HVDC Transmission

High voltage direct current (HVDC) systems are used for the following reasons:

1. When the power that need to be transferred is large.
2. When the distance from power stations to consumers is a long.
3. For interconnecting different grids.
4. In big cities DC lines relieve congestion or bypass complicated AC cabling systems.
5. In undersea cables, the charging of the capacitor like structure of these cables every half cycle is avoided. The undersea cable is a capacitor because there is a conductor, then XLPE then a sheath of steel. This is a longitudinal capacitor. HVDC need to charge up this capacitor once while HVAC need to charge it up every half cycle.
6. To enable the use of renewal energy. The output of solar is DC and wind turbines turn at different speeds so the output of one wind turbine cannot be joined to another. But if all the power of different wind turbines are converted to DC the output can be joined.

The conversion of power between AC and DC was enabled with the development of:

1. Mercury-arc valves before the 1970s
2. Thyristors especially Gate Turn-Off (GTO) thyristors after the 1970s.
3. Integrated Gate Commutated Thyristors (IGCTs)
4. MOS Controlled Thyristors (MCTs)
5. Insulated Gate Bipolar Transistors (IGBT)

1 to 4 are CSC (Current Source Converters) or LCC (Line Commutated Converters). (5) is a VSC (Voltage Source Converter). The demand for IGBT (or VSC) has increased dramatically because of the proliferation of off shore wind turbines installations. Imagine a flood control gate controlling the flow of a river. That is how a Gate (G) controls the flow of current from Source (S) to Drain (D) in an IGBT. In bipolar transistors, current must be sent to the Gate to enable current flow from Source to Drain (actually the names of the legs in BJT is Base, Collector and Emitter but it is just named that way it's function is same as G, S,D). But in IGBT, a voltage at the G induces and electric field which enables current to flow from S to D. Thus the categorization of IGBT as Voltage Source Converter (VSC). If current need to go into the G, it is categorized as Current Source Converters). The demand for various power electronic used in the electric power industry is shown in Fig. 100.

Power Electronics usage

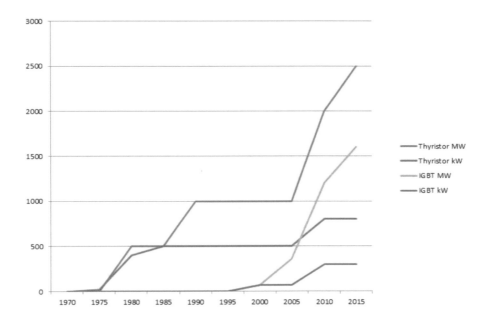

Fig. 100: Demand for various electronics in the electric power industry

Insulation thickness and conductor spacing of an electric cable is determined by the peak voltage. Thus for an AC cable carrying 275KV, the cable has to be able to withstand voltages of: . But a 275kV DC line needs to be able to handle only slightly above 275kV (catering for a slight surge). Which also means a 275kV AC cable can potentially carry 389kV of HVDC.

Due to all these factors, HVDC lines can carry **50% more power** than HVAC lines. The HVDV lines are also **30% cheaper** to construct than equivalent HVAC lines. The advantages of HVDC over HVAC systems for carrying large power include higher power ratings for a given line and better control of power flows.

The higher carrying capacity of HVDC is important because upgrading current lines or erecting new towers can cost billions of dollars. Many newly installed lines now use HVDC for long distance, high load transmission, especially in developing countries such as China, India and Brazil. In India existing HVAC lines have been used to carry HVDC thereby drastically increasing the current carrying capacity. Each cable can carry 50% more power with HVDC. In a typical HVAC line there is L1, L2, L3 (R, Y, B) bundled cables (two cables per phase) on the right of the tower which leads to 50% X 6 = 300% extra current carrying capacity. This is shown in Fig. 101. If the Single Wire Ground Return (SWGR or Single Wire Earth Return (SWER)) configuration is used there will be another six wires on the left side of the tower which need not be used as neutral, thereby increasing her current carrying capacity to 50% X 12 = 720%.

Fig. 101: HVAC bundled transmission lines on left versus HVDV bundled transmission lines on right. Only positive, negative and Ground for HVDC

Submarine HVDC

As an indication of the current carrying capacity of plastic covered cables versus bare cables the following statistics are useful:

1. The longest undersea cable (heavily covered by plastic) so far is **580 km, 450kV**, 700MW XLPE HVDC cable between Norway and Netherlands.
2. The longest bare overhead line is the **2,385 km, 600kV**, 7.1GW HVDC Rio Madeira transmission link in Brazil.
3. A close second is the **2,090 km, 800kV**, 7.2GW HVDC Jinping-Sunan transmission link in China.

Added to the above statistics, underground cables fail much more often than overhead cables and the overall cost of underground cables is up to 400% higher than bare overhead cables. Therefore submarine cables are bad compared to overhead lines but HVAC submarine cables are far worse than HVDC submarine cables.

Currently available undersea XLPE cable has lots of capacitance loss. Each conductor is surrounded by a relatively thin insulator (XLPE) and the final layer is a metal sheath. This conductor, the dielectric (XLPE) and the final metal sheath corresponds to a capacitor. Thus the whole HV undersea cable is a longitudinal capacitor. This capacitance is parallel with the load. When AC is transmitted through the cable, additional current must flow to charge the cable capacitance on each half cycle. This extra current flow results in energy loss via dissipation of heat; thereby raising the temperature of the undersea cable. This thereby causes the cable to weaken as a current carrying conductor. When DC is used, the cable capacitance is charged up only upon first energizing so the energy loss is less. This is why most undersea cables today are carrying HVDV (High Voltage Direct Current).

Another loss is dielectric loss. Here the varying AC electric field causes small realignment of weakly bonded molecules within the dielectric material causing small vibration of molecules, thus heating up the XLPE. If XLPE is used as the dielectric, this effect becomes significant beyond 63.5 kV in mineral (magnesium oxide) filled cables and 27 kV for non-mineral filled cables.

Therefore underground power transmission has a significantly higher cost and greater operational limitations but it's utilization is use is expanding rapidly mainly for esthetic reasons. There is also some advantage in right of way.

Disadvantage of HVDC

The disadvantages of HVDC are in conversion, switching, control, availability and maintenance. The cost for all these is much higher than for HVAC. With HVAC the only switching is for switching off or on a line. But for HVDC the switching need to happen continuously to create the DC. Most of these switching is with thyristors and IGBT. The main circuit breaker for HVDC is much more complicated. With the recent invention of HVDC switchgears by ABB, fast switching of HVDC line have been enabled. Switching of HVDC has always been tricky because it does not have the advantage of HVAC where the actual switching can be made to happen when the AC waves cross the zero volt region. With HVDC mechanism must be included in the circuit breaker to force current to zero, before the actual switching occurs. Otherwise too much arcing will occur wearing the silver plated copper contacts. In HVAC switching is performed when the AC wave crosses the V=0V so L1, L2, L3 (R, Y, B) cannot be switched off or on at the same time as shown by Fig. 102. The switching of the L1, L2, L3 (R, Y, B) phases must be done about 4ms (5ms British) from each other.

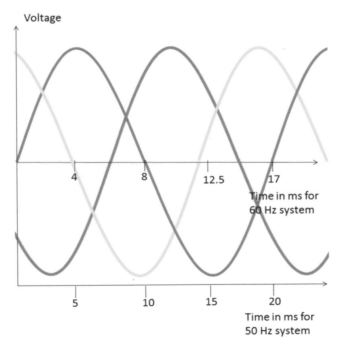

Fig. 102: Three phase waveform

The tyristors and IGBT also induce losses and they have limited overload capacity and are expensive. At smaller transmission distances the losses in these power electronics may be bigger than in an AC transmission line. The cost of the inverters may not be offset by reductions in line construction cost and lower line loss.

Chapter 25
Grid Control

Large National or International Grids

The solution for the intermittant nature of renewal power is to join up grids as much as possible. Thus a low wind flow over wond turbines in South Dakota can be backed up by gas turbines in Texas or solar farms in California.

So the North America has decided to join up it's nine major grids with Back-To-Back DC lines (B2B). These B2B can be the length of two classrooms to concert AC to DC at one end and DC back to AC at the other end. A large substation is being built in the U.S. called Tres Amigas which will interconnect the three large grids in the U.S, the Western Interconnection, the Eastern Interconnection and the Electric Reliability Council of Texas (ERCOT) grid. Electricity from each grid is converted to DC and flows in a smaller DC high-temperature superconductor (HTS) wires in a 58 km² area where any of the three grids can pull out power or supply intermittent renewal energy from wind or solar farms. Superconductors have the ability to keep electrons flowing within it for long periods of time making them energy storage devices. The Tres Amigas substation can initially handle 5 GW and is expandable to 30 GW. HTS wires conduct electricity about 200 times better than copper wires of the same dimensions.

In Europe the International Electrotechnical Commission (IEC) did research to determine how extensive usage of renewal energy can be achieved to prevent global warming and reduce dependence on fossil fuels. Their proposed design is a Europe wide grid to cater to the intermittent sources of electric power like wind or solar. This way, extra power from fossil fuel or nuclear power stations can back-up a sudden low wind situation around wind farms or sudden cloud-cover over solar farms. Therefore the Synchronous Grid of Continental Europe was built and it is currently the largest electrical grid in the world. The frequency of the grid is 50Hz and it serves over 400 million customers in 24 countries. Large energy storage has also been determined to be one of the solutions; batteries with liquid electrodes have been suggested as a form of grid energy storage. In Germany, solar panels power pumps water up a high tank and in the night water is released from these tanks to produce hydroelectricity.

Since computer simulation have shown that giant faults cannot pass a B2B system, a fault cannot bring down the whole North American power system but it may be possible that a major fault can bring down the whole European grid.

Control of Grids

In the vast grids the V and I tend to divert and not flow synchronously. This will cause the PF to increase and thereby energy loss. To control the drifting of the I wave from the V wave the following are used. To produce vars:
1. Generators run as Synchronous motors and over-exited
2. Capacitor banks
3. Switching out of reactors
4. The capacitance of overhead lines and underground cables

To consume vars the following are used:

1. Induction motors
2. Generators run as synchronous motors under-excited
3. Reactors in substations being switched on to the grid line as shown in Fig. 103.
4. Inductance of overhead lines
5. Transformer inductances
6. Line commutated static converters

Inductor, Capacitor, Resistor model of HV overhead lines

Overhead lines are modeled as a resistors and inductors (due to magnetic flux around the cables) in series with capacitors shunting the overhead lines to ground. Capacitance is negligible for line spans shorter than 50 miles (80 km).

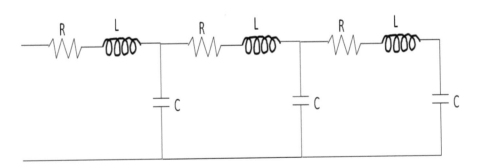

Fig. 103: Model of transmission line which depicts it as resistors in series with inductors with capacitors shunting to ground

Transmission line are both inductive (absorb vars) and capacitive (generate vars); meaning they absorbs and generates reactive power respectively. Under light load, power lines are more capacitive (generates vars) than it consumes (inductive).

When there is a large loading in the power lines it consumes and generates the same amount of reactive power.

But under heavy loads, the power lines absorb more reactive power than it generates. HV lines are spaced further apart to achieve sufficient insulation in-between them so they have higher inductance. What happens is when lines are further apart; the flux lines do not cancel out each other (superimpose with each other) as when they are closer together as in a three core, three phase underground cable. The flux lines do not cancel and there is no insulator to reduce the flow of flux lines which are not superimposed. Upon observing from one point of L1 conductor will cut a large length of L2 conductor with increasing concentric radius of flux lines. So effectively L2 conductor will experience more flux lines. Since number of flux lines is proportional to the inductance, the inductance is increased.

For underground cables the flux lines cancel out each other so the inductance in series is much smaller. Also the space in-between the phase conductors are very small resulting in much smaller number of flux lines crossing from one point of L1 to any length of L2 for example. In addition only a few flux lines that can pass through the XLPE in-between the power lines. But the capacitance in parallel is relatively very high because the three phase conductors with XLPE in-between is equal to a capacitor; the underground cable is effectively a longitudinal capacitor from energy intake to energy out-take.

For overhead lines, the greater the radius of the power lines, the lower it's inductance. But large radius power lines are hard to work with due to their inflexibility and weight. Thus two or more conductors are bundled together to approximate a large diameter conductor. This will also reduce radiative corona loss.

Since a voltage V is applied to a pair of conductors separated by a dielectric (air), charges q of equal magnitude but opposite sign will accumulate on the conductors. Capacitance C between the two conductors is defined by:

$$C = \frac{q}{V} \qquad (67)$$

The capacitance of a single-phase transmission line is given by:

$$C = \frac{2\pi\varepsilon}{\ln\left(\frac{D}{r}\right)} F/m \qquad (68)$$

Where ε = 8.85 X F/m

This capacitance is only significant beyond 80 KM.
Without balancing watt (active power) the frequency in the system will be affected. Without balancing vars voltage in the system will be affected.

Note production of vars is done with compensators:

1. Capacitor banks (Fig. 105)
2. SVR (Static var Regulator)
3. running a generator in synchro mode (as a motor)

4. Shunt reactors (Fig. 104)
5. Production of vars (increase capacitance) is by decreasing loads on lines (load shedding).

Number (3) is done by playing around with the DC going into the rotor of the generator. Slightly higher than optimal DC (when V and I are in phase) will cause the voltage wave to move slower (harder to move the rotor because it has become a stronger electromagnet). This will cause the current wave to move faster than the voltage wave (voltage wave follows rotor rotation) which is current leading, capacitive or var producing.

Consequently slightly less than optimal DC will cause the voltage wave to move faster (easier for the rotor to turn as it has become a weaker electromagnet). This means the current wave will lag the voltage wave which is current lagging, inductive or absorbing vars.

The production of reactive power should be as close as possible to the inductive loads because vars cannot be transmitted far over the line since X ≫ R in a grid. Factories and big buildings are required to have capacitor banks to produce vars to balance the vars consumed by their induction motors and other loads. Note humans only use inductive machines (motors etc.), this author have not come across a capacitive load.

The further a city is from the power station the more V and I wave move apart, meaning the lines have more vars. Thus compensators (1-5 above) need to be placed close to such a city. If microprocessor controlled thyristors are used to control the current flowing through a reactor-capacitor combination, a continuous control of the reactive power can be achieved; such a device is called SVC (Static Var Compensator).

For example for the city this author lives in, previously there was enough power generation in the city but currently the power generation in the city is only about a third of the demand. Most of the power is generated about 1000 miles away in a hydroelectric power station. Thus the phase shifting of V and I is quite high in the city. So two 26MW (out of four) hydroelectric generators about 100 miles away is run in synchro mode to compensate, but recently even this is insufficient so a SVR is being built just outside the city.

Fig. 104: Reactor Banks at a Substation, 33kV on left and 275kV on right

Fig. 105: Capacitor banks in Substation, 275 kV

Power Quality

Though the word power quality seems to indicate electrical power it is the voltage variation which need to be solved because the current demand by appliances cannot be controlled. Various categories of power quality defects are termed as swells, dips or sag, random, spikes or surges, under voltage, over voltage, brownouts, and harmonic.

Power quality has traditionally been affected by switching on or off (with switching off having more effect) of loads in the grid. Therefore a large factory switching big equipment will affect the whole grid. Starting large generators also affect the power quality, which are generally solved by AVR (Automatic Voltage Regulators) and other equipment at the power station.

Power quality has become important recently with the increase in sensitive electrical devices being built. Switch Mode Power Supplies (SMPS) has become the standard intake for all electrical devices from hand phones, televisions all the way to advanced equipment in factories. SMPS works with the principle of initial rectification of the power company's 60 Hz (50 Hz British) electrical waves with a bridge diode. The DC produced is then inverted into an ever increasing frequency AC wave (up to 1,000,000 Hz), stepped down with a transformer and rectified again to be used by devices. This seems to be an optimum solution. High frequency AC rectifies to a very flat DC and the transformer decreases in size with frequency, thereby saving copper.

But in reality if the system above is fed with an initial uptake of not perfect power company's 60 Hz (50 Hz British) but spiky waves with lots of noise or harmonics. The initial rectification system which is supposed to get rid of all variations does not clean the 'dirty' waves which do get through the initial rectification thereby destroying expensive equipment. For example a high spike on the positive side will enter the bridge diode capacitor system of the rectifier and remain a high positive all through the SMPS as a high frequency spike, thereby going into the load. Doctors in developing countries have long complained that expensive medical equipment imported from overseas tends to get damaged very fast. The same must be the case for all sensitive equipment users in industry of developing countries.

The perfect solution developed today is to use a UPS (uninterruptible power supply). In this case the power company's 60 Hz (50 Hz British) is stored in a battery and output from this battery is inverted into high quality 60 Hz (50 Hz British) supply. This is because only battery can supply a perfect DC.

One solution to power quality problems is to use a phase shifting transformers which have a non-integer or complex turn ratio such as $1: ej^{30}$. This can cause a phase shift of voltage and thereby absorb or produce reactance. Below are three more problems power quality problem causes and solutions to them.

Resonance between the capacitors and inductors in the grid. This can cause a high voltage surge which can breakdown insulation and therefore. The solution is to filter to remove the harmonics.

Travelling waves are a problem especially when high voltage waves reflect on reaching an end or junction of the grid. The affect is that these reflected waved get superimposed with the original AC wave creating a wave that has an amplitude that is several times the original AC wave. The solution to this problem is to have a proper switching sequence. A smarter grid that can detect any surge in amplitude is a fool proof solution. Reduction in sending HV over underground cable will help because the likelihood of air insulation breaking down for overhead cable is much less.

Sustained overvoltage which will affect frequency. If voltage is not controlled, that is supply and demand not balanced overvoltage can occur. An overvoltage is caused by supply greater than demand, consequently over-demand will cause a low voltage. An overvoltage will result in the rotor of the generator turning faster and thus the frequency increasing. Note it is easier to remember that as more than normal current is drawn from the stator coils, it becomes a stronger electromagnet which will slow the rotor thereby reducing the frequency of the AC wave. So in this case there is an over-supply suddenly so the amps drawn will be less than normal making the stator coils less magnetic so the rotor will turn faster. Such a scenario happens due to poor voltage control at the control room. Smarter grid to detect and eventually automatically control in the control room is a solution to this problem.

Smart Grids

Smart grids is a grid with intelligence built in. This will enable the full control of the grid. Smart meters have also been deployed to a high degree in USA. These meters can send and receive information from utilities and communicate wirelessly with appliances especially the cloth washing machine which is one of the biggest consumers of electric power in USA homes especially because of their dryers. A customer can switch on the washing machine but it will not switch on till the load demand in the region is low; normally around 2-3am. Eventually more and more of the appliances in the homes can be monitored which gives the power company an idea of how much inductive or capacitive load out there. Electric meter readings will be sent directly to the power stations. There are also developments to use the internet protocol (TCP/IP) which is very much faster than the currently used Ethernet (mostly with SCADA). Currently most power companies are worried about putting all their equipment on the internet because of the possible hacker attack but the Ethernet can just as easily be hacked. As virus protection develops software is developed, the firewall can be so high that it is a Himalayan task to launch an attack. Gone are the days when an individual can launch an attack, today it takes a large country to launch a virus attack. With the TCP/IP carrying all the faults response time can be faster for sudden faults. Sudden surge in load and insufficient generation can be rectified faster. Generators can be started automatically instead of the manual system currently practiced. Basically the Smart Grid utilizes the advances made in computer and communications technology into the grid. Among the various devices developed are sensors with built in microprocessors which can take action instead of just detecting faults. GE is working on a digital twin for all future gas turbines it produces. With sensors installed in all future GE GTs, when a bearing in a GE GT in Malaysia fails, it will show up in a computer running in a cloud in USA.

ABOUT THE AUTHOR

Dr. Prashobh Karunakaran is a Senior Lecturer in Electrical Engineering at Universiti Malaysia Sarawak (UNIMAS), Malaysia. He is a professional engineer and also runs an electrical consulting and contracting business together with his electrical technical training school. He did his Bachelors and Masters at South Dakota State University, SD, USA and his PhD at Universiti Malaysia Sarawak (UNIMAS). His wife, Sreeja is also an electrical engineer and they have three children, Prashanth, Shanthi and Arjun.

Dr. Prashobh Karunakaran can be reached at www.unimas,my.

Printed in the United States
By Bookmasters